# COVID-19

# COVID-19

## THE PANDEMIC THAT NEVER SHOULD HAVE HAPPENED AND HOW TO STOP THE NEXT ONE

### DEBORA MACKENZIE

hachette
BOOKS

New York

Hachette Books
Hachette Book Group
1290 Avenue of the Americas
New York, NY 10104
HachetteBooks.com

Twitter.com/HachetteBooks

Instagram.com/HachetteBooks

First Edition: June 2020

Published by Hachette Books, an imprint of Perseus Books, LLC, a subsidiary of Hachette Book Group, Inc. The Hachette Books name and logo is a trademark of the Hachette Book Group.

The Hachette Speakers Bureau provides a wide range of authors for speaking events.

To find out more, go to www.hachettespeakersbureau.com or call (866) 376-6591.

The publisher is not responsible for websites (or their content) that are not owned by the publisher.

Print book interior design by Six Red Marbles, Inc.

Library of Congress Control Number: 2020938926

ISBNs: 978-0-306-92424-8 (hardcover), 978-0-306-92423-1 (ebook)

Printed in the United States of America

LSC-C

10 9 8 7 6 5 4 3 2 1

*For James, Jessica and Rebecca, who make everything possible.*
*And in grateful recognition of the scientists and journalists who do*
*their best to find out what's going on and try to save us from it.*

# CONTENTS

# PREFACE

In November 2019, a coronavirus from a common little bat jumped, somehow, to a human, or maybe a few of them. It just happened that the virus could spread easily among people already, or it evolved fast, as these viruses can. By December, a cluster of people were hospitalized with severe pneumonia in Wuhan, China, and it wasn't the flu.

Not enough was done to contain this new virus until January 20th, when China told the world it was contagious. By then there were already so many cases in Wuhan, the city had to be locked down three days later to control the epidemic—but it had long since spread all over China and to other countries. The virus was named SARS-CoV-2 because it was so similar to another one we had barely managed to beat back in 2003. As you know, the disease it causes was named Covid-19: "co" for corona, "vi" for virus, "d" for disease," and 19 for the year it appeared. A lot of people just call it the coronavirus.

Three months after Wuhan was locked down, some two billion people worldwide were also in some form of lockdown, and everyone, everywhere faced infection with the virus, with few effective treatments and no prospect of a vaccine anytime soon.

Covid-19 has infected the entire human world. This pandemic has been like a big dog, picking up our fragile, complex society in its teeth and shaking it. Lots of us have died. Lots of us will continue to die, either from the virus itself or from the long-term poverty, political and economic dislocation, and overloaded medical systems that will be the pandemic's legacy. Some aspects of our society will change for the worse, some maybe for the better—but, either way, for good.

And through it all, we have been deluged with reams of news reports and instant analyses, heartbreaking frontline accounts, revised government instructions, and new medical advice, plus probably the most staggering global outpouring of instant scientific research in history, all trying to predict what's coming next and figure out how to mitigate this disease disaster.

But you know all that.

And still there is the question: How could this happen? This is the 21st century. In much of the world, we have wonder drugs and flush toilets and computers and international cooperation. We don't die of pestilence anymore.

Sadly, as we all know now, yes, we do. But what is especially sad for a science journalist like me who writes about disease for a living is that this pandemic has not exactly been a surprise. Scientists have been warning for decades, with mounting urgency, that this was going to happen. And journalists like me have been relaying their warnings that a pandemic is coming and that we aren't prepared.

How did we find ourselves in this situation? In short, there are more and more people, and too many of them have had to put ever-increasing pressure on natural systems to get the food and jobs and living space they need. That means pushing into wilderness that harbors new infections and intensifying food production in ways that can breed disease. Covid-19, Ebola, and worse come from destroying forests. Worrying flu strains and antibiotic-resistant bacteria come from livestock. Yet we have neglected to invest in the things that discourage infectious disease: public health, decent jobs and housing, education, sanitation.

Then the impact of the new pathogens we unearth is magnified by our ever-increasing global connectedness, as we crowd into cities and trade and travel in an ever-denser global network of contact. So once public health fails and contagion appears anywhere, it goes everywhere. We know so much about beating disease, yet fragmented governing structures, lack of global accountability, and persistent poverty in so many places ensure that those failures happen and disease propagates.

Despite all that, we know what we need: much better understanding of potentially pandemic infections, fast detection of new outbreaks, and ways to respond to them quickly. I'll be looking at that in this book. So far, we haven't been able to do that effectively, where it is most needed.

In 2013, two labs—one Chinese, one American—investigated a tribe of bat viruses that are almost certainly the source of Covid-19. They immediately recognized the threat. One lab called them "pre-pandemic" and a "threat for future emergence in human populations." The other wrote that they "remain a substantial global threat to public health."

Nothing was done. We could have learned more about them,

designed some vaccines, looked into tests and treatment, studied ways these viruses might infect human populations—and shut those down. None of that happened. It was no one's job to take on those tasks with this kind of threat, even when it materialized.

Yet we needed so much to be in place if one of these viruses went global—which one did. You don't need to be told. Testing. Ventilators. Drugs. Vaccines. Protective gear for doctors and nurses. A plan for using old-fashioned quarantine and isolation to stop this kind of virus from spreading. A plan for dealing with the economic impact. Measures to contain the virus so we might not even need those things. Experts and governments have been talking intensively about pandemic preparation for nearly two decades, and still we weren't prepared.

And this kind of virus wasn't—and isn't—even the only viral threat out there, yet we're just as unprepared for the others. I wrote the following for *New Scientist* magazine in 2013, the year the Covid-like viruses were discovered, about a visit to the World Health Organization's then-shiny new situation room and what might happen if H7N9 bird flu, the virus causing concern at the time, went pandemic:

> As it stands, the World Health Organization's top brass will watch any H7N9 pandemic unfold from their strategic operations centre. Information will flood in; body counts will mount. Governments will be told that their demands for vaccines and drugs cannot be met. They will issue declarations, hold briefings, organise research, tell people to wash their hands and stay home. Mostly, though, they will just watch helplessly.

Sound familiar? Especially the part about washing your hands and staying home?

I don't claim to be prophetic: I'm not. Other journalists and scientists have said as much and more. As far back as 1992, the top infectious disease scientists in the US warned about "emerging infections," declaring that the threat from "disease-causing microbes…will continue, and may even intensify in coming years."

If that sounds like unusually cautious language, even from scientists, it's because they were afraid any stronger language would trigger disbelief. That's almost all that has changed.

It's not that they weren't heard. In the years since then, we all started half-expecting a pandemic. Pandemics became part of the cultural background noise, reflected, with varying balances of science and entertainment (and zombies), in films like *Outbreak*, *Contagion*, and *I Am Legend*. There was some disease surveillance set up, new international rules written, a lot of virus research. A few countries had pandemic plans, on paper. Yet when the lockdowns began, in many places toilet paper was in more demand.

The only real surprise when Covid-19 finally hit was the sheer extent to which most governments simply had not listened to the warnings. We were unable as a planet to muster our considerable scientific understanding of disease in time to soften the blow, never mind preventing it in the first place. And, as I will explain in the coming pages, we could have—at least a lot more than we did. Science didn't actually fail us. The ability of governments to act on it, together, did.

Experts had warned about the lack of preparation in addition to the risk of a pandemic itself. The few countries with pandemic plans built them around a very different virus, flu, and regardless, many failed to stockpile or acquire the most basic essentials for making the plans work. I'm not sure their response would have been much more effective if this had been a flu pandemic. Which we will have at some point.

The World Health Organization (WHO) made it very clear how to contain Covid-19, but few countries followed their advice entirely. A few showed what should have been possible for all countries. The rest did pick-and-choose variations on the WHO's advice and/or that of their scientific or political advisers. Nearly all countries were more or less too late to limit the damage as much as they might have, and the pain of lockdowns and economic dislocation in some places seemed to rival the disease.

But you know that.

So, besides the question of how this could happen, the other big questions are: Can this happen again? And can we do better next time? The answer to both is yes. Some real pandemic planning is now in order, because the Covid-19 pandemic may not even be the worst we could see. And even Covid-19 could still have some tricks up its tiny sleeves.

First, let's look at the immediate future from the virus's perspective.

Eventually, after considerable death and disruption, most people in the world will have been exposed to or vaccinated against Covid-19 and will be, we hope, immune to further infection with the same virus as a result, at least temporarily. So, with fewer people around that it can still infect, new cases should slow to a trickle. The virus might even quietly die out, as its sister-virus SARS did in 2003 when we blocked enough chances for it to spread.

Or it might adapt to its new situation. RNA viruses like this one can evolve quickly, although the Covid-19 virus isn't quite as volatile as some. Like flu, it might mutate to evade the immune defenses our bodies will eventually learn to mount and start another global rampage, perhaps a bit less deadly this time—or perhaps a bit more. The comforting myth that viruses become more benign as they adapt to us is simply not true. It all depends on what works for the virus, and it can go either way. We will look at that later in the book.

Or it might circulate and surge sporadically, perhaps pouncing on new, susceptible humans, becoming yet another disease of childhood.

This pandemic has moved fast since it started. You may already know something about which of those scenarios is playing out. There aren't, broadly speaking, a lot of different things a disease can do, bound by the implacably quantitative laws of epidemiology, the science of epidemics.

Until then, horrific as it has sometimes been, we can be grateful it hasn't been worse. Covid-19 does not have a massive death rate—best guesses as I write this are that it's less deadly than we initially feared, but still maybe ten times more deadly than ordinary flu. SARS was ten times deadlier than that. Fortunately, it never learned to spread like Covid-19—and, with luck, Covid-19 will never learn to kill like SARS. Think about what this pandemic would have been like with ten times the death rate.

And as many of us have painfully learned, Covid-19 mostly kills older people. Speaking as one myself, I don't wish to be cavalier about this, but the brutal fact is that losing people in old age does not cause as much economic or social disruption as losing people of working and childbearing age. And even that will pass: in a year or three, with luck, we may have drugs and vaccines to protect everyone, including the elderly.

So why write a book about this when there's still a lot we don't know? Because we already know enough to say some important things, and we need to do that while memories of these hard times are raw enough for people to hear them.

The first thing to say is that this was predicted and could have been, to a large extent, prevented.

As for prediction, I am just one of many journalists who has been warning about the threat of a pandemic since the 1990s— and some were at it earlier. Since at least 2008 the US Director of National Intelligence has warned the president that a pandemic of a virulent, novel respiratory virus was the most serious threat the country faced. In 2014 the World Bank and the OECD, the club of rich nations, called a pandemic the top catastrophic risk, out-weighing terrorism. Bill Gates has been warning that we aren't ready for a pandemic for years.

Second, this pandemic won't be the last one. There are simply too many potentially pandemic germs out there to predict which will emerge next. But before Covid-19 happened, we knew corona-viruses were among the leading possibilities: they were on a WHO watch list. Even with such warnings, we didn't do enough prepara-tory work on drugs and vaccines for coronaviruses like Covid-19 to allow us to easily adapt and produce them now—and we still haven't for many other viruses that pose a threat, including H7N9 and its kin. We need to do that now.

We also need to do some serious pandemic planning for when the next one happens. The Center for Health Security at the Johns

Hopkins Bloomberg School of Public Health was among the institutions already trying to do that. Among other efforts, they were running computer simulations of hypothetical pandemics as a training exercise for public officials. A month before the first cases appeared in Wuhan, they ran one called Event 201, starring a fictitious virus that was nearly a dead ringer for Covid-19. I can think of few better illustrations of how we knew this was coming.

Let me emphasize that this was a total coincidence: this was a "what if" scenario playing out in a computer model of US society, featuring a made-up virus. They chose a coronavirus for the simulation partly to show how disruptive even a relatively mild virus can be.

They succeeded. The result of the simulation was what we are living out now: overwhelmed health care, disrupted global supply chains, needless death, economic dislocation. And a table full of officials from government and industry sitting there saying, *If this were to happen, there's not much my sector/department/office could do.*

And the people who wrote that simulation were going easy on the officials—maybe so they'd sit through the entire afternoon and not be so horrified they'd quietly slip out at the coffee break, trying to forget what they'd seen. There are much worse viruses out there that could trigger a pandemic, that would kill more people, at younger ages.

It will not be much comfort to those who have lost or will lose loved ones to Covid-19, but so far, believe it or not, we've been lucky.

In addition, what almost no one realized before Covid-19 happened—I don't know how many realize it now—was what a pandemic could do to our complex, just-in-time society, and that economic domino effects would cascade through our tightly coupled global support networks.

What we need to remember, though, is that we *will* have another pandemic. And it could be worse.

So we have to do better—and we can. The hard-earned good news is that Covid-19 has shown us what we need to do. We cannot let a virus catch our interconnected global community this stupidly flat-footed again. We cannot let it break those interconnections either, at least not all of them. If this pandemic teaches us anything, it is that up against a contagious disease, we are all in this together. One big early lesson was that no country can really seal off their borders anymore, or go it alone. Our society is global; our risk is global; our response and our cooperation must be global.

I can't think of a time when this pandemic will be "over" enough to provide a better vantage point for looking at these things. When the virus does grind to a halt, or we tame it with vaccines, it seems all too likely that we will drift back into a status quo of spending on wars and weapons—and on recovering from the economic damage Covid-19 is doing—not on preparing for the next virus. We will need to forget this nightmare, and to judge from past pandemics, we will.

Yet at this moment, the subject has our undivided attention. We can already say a bit about how this happened, and why, and what our options are to start doing better. Many scientists know this, and governments, we hope, will learn. But a lot of other people need to think about this too, whatever you do in life, in the kind of detail that will allow you to help make the changes we need.

In any disease emergency, certainly in a pandemic, it is vitally important to tell everyone the whole truth—what we know and what we cannot know—and not hold back for fear of scaring people. That is a mistake that governments and other authorities make repeatedly with bad news stories like diseases.

What is happening might be scary, but saying so might galvanize people to take more effective action. Sometimes fear is necessary. That's why we have it.

But it shouldn't have to come to that. This is where you come in. Learning from this pandemic and preventing the next one will take political action of all kinds, from everyone.

The more people understand what we need to do, the more likely it is to be done. People vote. People march. People pressure. People decide to study virology or public health or nursing or vaccine engineering or communications. Public activism drove the development of HIV drugs and made them affordable. It drove the introduction of sanitation, the massive success of vaccination, the beginning of the end of smoking.

We can do it again. We have to.

To find out what is happening with Covid-19 right now, read the news. For exposés and analyses of what this and that government or politician did wrong in dealing with it, also read the news and the stories that will pour out over the coming years. I know I will.

In this book, I'm going to give you the big picture. We'll take an in-depth look at what happened and whether we could have stopped it, before looking at the recent past to learn the natural history of some of the more amazing natural phenomena that make us deathly ill. We'll see how previous pandemics and threats of pandemics should have prepared us and learn the lessons we failed to apply before and after Covid-19 emerged. Then we can talk about what we need to do better before the next one hits.

I hope that, eventually, we will do more than talk.

## A NOTE ON THE VIRUS

The WHO in its wisdom decreed that this disease would be known by the unlovely name Covid-19. Many people, and many languages, have just stuck with calling it the coronavirus. That is a much prettier word, but strictly speaking, it refers to a whole family of viruses, to which the one that causes Covid-19 belongs. I will be using the word *coronavirus* for that family.

The virus is officially called SARS-CoV-2, a name chosen by a committee of virologists expressly to underscore how novel it isn't and how similar it is to the virus that caused the disease SARS in 2003. That virus was renamed SARS-CoV-1. That makes the official name confusing, so I hope the virologists won't mind if I try and call it the virus that causes Covid-19, or even the Covid-19 virus, where at all possible. Because it is, and the word doesn't immediately hit the unspecialized eye as referring to another disease.

## CHAPTER 1

# Could We Have Stopped This Whole Thing at the Start?

> "Every disaster movie starts with someone
> ignoring a scientist."
> —Popular poster at the April 2017
> March for Science.

So how did we end up with the Covid-19 pandemic? Could we have stopped it once it started? Could we have stopped it from starting at all?

If your house burns down, you ask two things. First, how did a fire get started in the house to begin with? Second, and most urgently, given that it did—and we saw it happen—why didn't we put it out before it spread? We'll look at the first question later in the book. Let's look at the second now. What happened to unleash a Covid-19 pandemic on the world?

The first inkling I, like many others, had of the gathering storm that became Covid-19 was a post on the online forum, ProMED. The machine-translated report from Finance Sina, a Chinese online news site, read:

> On the evening of [30 Dec 2019], an 'urgent notice on the treatment of pneumonia of unknown cause' was issued, which was widely distributed on the Internet by the red-headed document of the Medical Administration and Medical Administration of Wuhan Municipal Health Committee.

It was December 31st, and in our suburban French village, just over the border from Geneva, the sun was coming up. I had family in for the holidays and had solemnly promised to stop working.

But, I told myself, that didn't mean I couldn't take a peek at ProMED, just to make sure I didn't miss anything important.

ProMED—the PROgram for Monitoring Emerging Diseases of the International Society for Infectious Diseases, a scientists' organization, formally called ProMED-Mail—is the world's leading online reporting system for new, or "emerging," infectious disease. Despite its importance, it's a non-profit run mostly by volunteer work, on a shoestring, with grants and donations. It was set up in 1994, as infectious disease specialists shaken by the emergence of AIDS in the 1980s uneasily realized that other new diseases might be out there, and that we needed an early warning system.

It consists of moderated daily reports of worrying medical events from contributors everywhere: doctors, vets, farmers, researchers, ordinary citizens, even agriculture labs (crops get diseases too). It's all in understated sans serif plain text—old-fashioned Helvetica, direct and to the point like the scientists who mostly read and contribute to it. Everything is classified by disease and place and date. The moderators, most of them veterans in their areas, tell you what they make of the reports, and I often cut straight to their comments. ProMED is one of the things humanity did right to prepare for disease emergencies like Covid-19.

For disease researchers, public health people, and science reporters like me—as well as anyone fascinated by the daily reality show—ProMED is required reading. When I ducked into my office that day, hoping it was early enough that my family wouldn't notice, the giant Sina Corp's financial bulletin was reporting people with severe, undiagnosed pneumonia in the central Chinese city of Wuhan, in Hubei province.

Many had connections to a seafood market. There were already 27 cases.

A red-topped bulletin—rendered red-headed by the machine translation—must be an emergency alert, I guessed. The reporter from Finance Sina had verified it by calling the official hotline of Wuhan's Municipal Health Committee the next morning. It was true. The story went out.

And it was worrying enough to make someone send it to ProMED. It wasn't hard to see why.

Pneumonia is not a disease caused by a specific germ, like measles or flu. It just means any infection that inflames your deep lungs, the part with the air sacs, called alveoli. Those sacs are what your lungs are all about: you suck air into them, and oxygen pours across the alveoli membranes into the oxygen-starved blood on the other side. The carbon dioxide waste in that blood meanwhile pours into the alveoli, and you exhale it.

If those delicate membranes are damaged by an infection, they can start leaking fluid, and the sacs can fill up. That stops oxygen from getting to the membranes and entering your blood. If this gets bad enough, you effectively drown in your own fluids.

A respiratory infection—be it virus, bacteria, or fungi—may invade your nose, throat, or the deeper, bronchial air passages and give you a cold or a bad cough. But if it gets into the alveoli, that's pneumonia, and it can kill you.

The fact that this pneumonia was undiagnosed was the red flag that got ProMED's attention. Normally, white blood cells defend your alveoli from the bacteria that are always there, pulled in by the billion on every breath. Winter flu viruses knock out this key part of our immune system, and then the bacteria can grow, causing pneumonia. Therefore, most winter pneumonia is first treated with antibiotics, which kill bacteria. In Wuhan, this apparently wasn't working. Nor, presumably, were diagnostic tests for flu or the other usual suspects.

The Municipal Health Commission was holding a special meeting, said the report. But they made a point of saying they thought it wasn't the SARS virus. SARS emerged in China in 2002 and rampaged through 29 countries in 2003, causing severe pneumonia and killing 774 people.

Good, I remember thinking. SARS may not get talked about much anymore outside the countries that were affected, except by us disease buffs, but it was vicious, packing a 10 percent death rate. It was stamped out with an enormous international effort—and luck—with only the classic techniques of isolation and quarantine, mainly because it was clumsy at spreading among people. But if this new illness wasn't SARS, what was it?

The market connection was worrying. A seafood market in China is also a "wet" market that sells live animals, and many sell exotic, wild creatures. The SARS virus came from bats and is thought to have jumped to people in a wet market.

Granted, there have been other reports like this on ProMED. In 2013, there was undiagnosed viral pneumonia in health workers in Anhui province in China. In 2006, people in Hong Kong had undiagnosed pneumonia after visiting several parts of mainland China. The ProMED moderator asked for more information in both cases, but further posts never appeared, so presumably no notable disease resulted.

This time, though, there was a worrying comment by Marjorie Pollack at the bottom of the post. Pollack is a doctor and epidemiologist, a 30-year veteran of the US Centers for Disease Control and Prevention (CDC), and the doyenne of ProMED's international team of moderators. She was involved in one of its proudest moments: alerting the world, on February 10th, 2003, to the mystery pneumonia in Guangdong later named SARS, nearly two months before China opened up about it.

What she wrote that holiday morning gave me that queasiness you get when you're trying very hard to dismiss a feeling of foreboding. Besides the news report, she observed, there was a lot of online comment about this.

Twitter and its Chinese counterpart, Weibo, weren't around when SARS broke out, but online chatrooms were. "The type of social media activity that is now surrounding this event is very reminiscent of the original 'rumors' that accompanied the SARS-CoV outbreak," wrote Pollack. "More information on this outbreak...

5

would be greatly appreciated. And," she added hopefully, "if results of testing are released."

What was different from SARS, she noted, was the transparency of the Chinese authorities. In February 2003, Chinese officials discouraged press reports on the undiagnosed pneumonia and did not immediately report it to the World Health Organization (WHO). They didn't start fully reporting cases until that April, by which time SARS had spread across China and East Asia and to Canada.

In the ensuing 17 years, there has been an astonishing revolution in China's politics and prosperity, so this new outbreak was occurring under very different circumstances. Chinese authorities told the WHO about it on December 31st. It later emerged that the first case occurred in November, but a respiratory infection during flu season had not seemed unusual, until hospitals started getting an unusual cluster of severe cases. The day after that, New Year's Day, the seafood market, which, it turned out, did sell wildlife, was closed.

But by January 3rd, Pollack still had no test results. There were worrying reports that people had been arrested for discussing online whether the mystery pneumonia might be a resurgence of SARS. Hubei authorities were cited as saying this was not true, as "no person-to-person transmission has been found so far."

That last part became a recurring theme. On January 8th, ProMED reported that China's Center for Disease Control and Prevention (China CDC) had identified the infection as a coronavirus, the same family of viruses as SARS, but repeated that there was no person-to-person spread.

I hadn't planned to go back to work yet, but I wondered if I should start looking into this story. It seemed unlikely to be significant without person-to-person spread. Animal viruses sometimes manage to jump to people, and even kill them, but fail to transmit between humans, like the notorious H5N1 bird flu. Without that, this outbreak might just fade, I thought hopefully.

But on ProMED, Pollack was sounding increasingly suspicious. So was Jeremy Farrar, the head of the medical research foundation The Wellcome Trust, and before that, head of Oxford University's medical research lab in Vietnam, where he dealt with SARS and H5N1 imported from China. On January 10th, he tweeted that if "critical public health information is not being shared immediately with @WHO—something is very wrong."

Something was. According to later press reports, doctors in Wuhan sent the public health lab at Fudan University in Shanghai a sample of the virus from a 41-year-old man who had been hospitalized with pneumonia on December 26th. He was a vendor at the now-closed Huanan Wholesale Seafood Market and became severely ill.

The Shanghai lab had it sequenced by January 5th. Unknown to them, the China CDC already had a sequence, but they had not made it public. Later, the Shanghai lab told Hong Kong reporters that what they discovered made them immediately contact the Wuhan health authorities and warn them to take action. The virus was from the same family of bat viruses that had spawned SARS.

On January 7th, China announced a coronavirus was causing the pneumonia. But when no further action was taken, the Shanghai lab posted the sequence on a public database, the first sequence published for the virus. China CDC then posted its sequence. The Shanghai lab was shut by authorities the next day.

The sequences allowed other labs to design specific tests for the virus. Other countries started screening travelers from Wuhan—and finding infected people.

Neil Ferguson and his team at Imperial College London are among the world's most respected mathematical epidemiologists: they construct complex, mathematical computer models that describe how diseases are observed to behave and then use them to predict how new ones will spread. In January, they used a large database of airline passenger statistics to calculate how many people in the catchment area around Wuhan typically travel internationally.

It stood to reason that the percentage of travelers who were found to be infected should be the same as or less than the percentage of the population back home that was infected, as there was no reason to think people with the virus would be more likely to fly abroad than people without it. But in fact, the percentage of travelers who were infected was much higher.

So, they inferred, there must be more infected people in the Wuhan region than was being reported. Imperial crunched the numbers—it's more complicated than simple percentages—and reported on January 17th that there were probably 1,723 cases, give or take, in Wuhan. Wuhan was officially reporting 41.

There was no need to suspect underreporting. The most likely explanation was more straightforward: official numbers counted only people who had a positive test for the virus, and in the early days of the epidemic, the only people being tested were those sick enough to go to a hospital. Other countries, however, were testing every traveler with a fever who had just been in Wuhan, even if they were only mildly ill.

The missing cases might simply have been not serious enough to go to the hospital. They would not, after all, have excited suspicion: mild cases look like flu, and it was flu season.

Still, looking at Ferguson's numbers, that seemed like a lot of cases for a virus that wasn't transmitted person-to-person. Or as the Imperial team drily put it, "Past experience with SARS and MERS-CoV outbreaks of similar scale suggests currently self-sustaining human-to-human transmission should not be ruled out." MERS, a virus with an even higher kill rate than SARS—around 40 percent—jumped to humans in 2012 and is, like SARS, a close relative of Covid-19.

Yet the official word was still that human-to-human transmission was limited at best. On January 10th, researchers at the University of Hong Kong found a family over the border in Shenzhen that became infected when they traveled to Wuhan. As the team published later, one family member did not go to Wuhan but became infected after the others came home. And doctors in Wuhan had also seen the sickness spread in families.

The researchers must have shared this information. On January 15th, Japan reported a case in Kanazawa who had just been in China but had not visited a wet market. The report noted that, according to the WHO, "there are currently cases in which the possibility of limited human-to-human transmission of this disease, including among families, cannot be ruled out. However, there is no clear evidence of sustained human-to-human transmission."

Sometimes, viruses new to people can jump to one or two more people, but get no farther: MERS does that.

On January 18th, the Wuhan neighborhood of Baibuting staged a potluck dinner with 40,000 people in honor of the kitchen god—and in a bid for a Guinness World Record for the number of dishes served. The mayor of Wuhan told a television interviewer later, after gatherings of people in Wuhan had been banned, that the party was allowed because they still thought that human-to-human spread was limited.

Then a local case turned up in Thailand. "Sticking my neck out to the chopping block, I suspect there may already be significant ongoing transmission of this novel coronavirus," Pollack wrote on ProMED. However, most cases were not being reported, because they were mild and unrecognized. About that, she wrote, "I obviously hope I am correct here."

By January 20th, cases were being reported across China, Japan, Thailand, and South Korea. Pollack's gloves were off. "It is becoming more difficult to conclude," she wrote testily, "that there has been limited person-to-person transmission as the case numbers are climbing."

Chinese scientists were also losing patience. Also on January 20th, Yi Guan, a virologist at the University of Hong Kong who helped uncover the SARS virus, told the Chinese magazine *Caixin* that the Wuhan outbreak was behaving like SARS: it was spreading between people.

The same day, China's president Xi Jinping finally went public, telling people to take measures to stop the virus spreading during the coming Lunar New Year holiday. Zhong Nanshan, an epidemiologist called the "hero of SARS" for helping discover the SARS virus in 2003 (and then telling the public it was out of control, when Beijing said otherwise), was heading the government's investigation. After Xi spoke, Zhong told China Central Television that the virus spread from person to person.

There were more surprises: the *South China Morning Post* in Hong Kong later reported that, according to classified documents they saw, the earliest case developed symptoms on November 17th, not December 1st, as later reported. It had taken China a month and a half to spot a problem and tell the WHO. The doctors involved knew it was contagious: early patients were put in isolation, and Zhang Jixian, head of respiratory and critical care at Hubei Provincial Hospital, told reporters in February she knew on December 26th when three members of one family had pneumonia. She made staff wear N95 masks.

What happened next shows how bad things already were in Wuhan by late January. To understand that, we have to look at the main ways of fighting an epidemic when you don't have drugs and vaccines: containment and mitigation.

Containment is by far the most effective way to limit an epidemic, if you get to it before there is a large number of cases. The classic method of epidemic control used for centuries is to isolate people with symptoms and then quarantine their contacts for the time it takes for them to incubate the infection and start showing symptoms. Maybe they won't have it—good. But if they do, the quarantine ensures they don't pass it on.

Nowadays, you can test people for the pathogen and only

quarantine those who test positive—if you really trust your test not to produce false-negative results. In either case, the chain of transmission is broken. Do that enough, and you can snuff out a virus: that's how the world defeated SARS.

However, this won't entirely work if the virus can spread before people show symptoms, as neither the person infected nor the people they contact will suspect a problem. And it is hard to do if more than a few people are sick. You have to trace and quarantine all the people each case might have infected, which can add up quickly with a virus as easily transmitted as Covid-19. You won't get everyone, so some new cases will continue to crop up, meaning more people to trace.

It's hard work. As it wrestled the Covid-19 epidemic to a halt in spring 2020, China eventually used six-person teams for each case to track contacts. The European Centre for Disease Prevention and Control estimates it takes a hundred person-hours to track one case's contacts. If you can break all the chains of infection from every case, the disease can be contained.

But you have to start early, before there are too many cases to track. If a disease is spreading generally—"in the community"—it becomes impossible: not only are there probably too many cases, but people might have no idea who they caught the virus from. That person could still be out there, spreading the virus, no matter how many known contacts of that case you quarantine.

At that point, the classic approach is to switch to mitigation. A lot of us know about that now because, with a few notable exceptions, most countries outside China didn't act in time to contain the virus and ended up mitigating: you ban large gatherings, close schools and workplaces, and generally reduce interaction between people to slow the spread of the disease, a set of measures known as social distancing.

At the extreme, as so many of us now know, you lock down and keep people inside. You don't entirely stop the spread of the virus, but at least it doesn't happen so fast that the sick overwhelm your hospitals. That means the number of cases you get per day or per week does not rise as high or as fast—the now-famous "flattening the curve." And even though you are in theory only slowing the spread, you also save lives, as more people who need intensive care can get it.

In the course of the Covid-19 epidemic, China discovered that outside Wuhan and Hubei province, a mix of mitigation and containment actually worked best: first contact tracing and quarantine to break chains of infection and then, if necessary, varying levels of mitigation to slow the spread of the virus, which, because fewer people were catching the virus from each person who had it, also made containment more feasible.

But on January 22nd, Wuhan was already at the point where lockdown was deemed necessary. To get to that point, there must have been considerable person-to-person spread. But with the official story being that the virus did not spread from person to person, officials could not make any visible efforts to isolate cases and trace contacts, back when it might have been possible to contain the virus. Now it wasn't.

As a result, China imposed a *cordon sanitaire* around Wuhan, a term from pre-vaccine days meaning "health barrier." They were invented for cities with the plague, so no one would enter—or escape, carrying the disease. English uses the French term because in 1821, France revived the concept, by sending 30,000 troops to seal the Spanish border to keep out the yellow fever raging in Barcelona.

No one could enter or leave Wuhan, a city of 11 million, without special permission, beginning 10:00 AM local time on January 23rd. That was extended to all of Hubei province a day later. Transport within the city was shut down.

But there was a huge problem: Lunar New Year was only three days away. This is China's biggest yearly celebration, when 400 million people travel to family celebrations all over the country—the biggest human migration on earth. Moreover, Wuhan is a hub for travel within China. Mass travel had already begun, and at news of the impending shutdown, people flooded into train stations and airports.

Authorities later announced that five million people had left the city before the *cordon sanitaire* could be enforced. Chris Dye of the University of Oxford and colleagues confirmed, using geographically coded mobile phone data, that 4.3 million people left Wuhan between January 11th and the start of the travel ban on the 23rd.

Many were carrying the virus. There was no way to call it back.

Back in Europe, my visitors had gone home, and I was now visiting family in London, with plans to hit the city's New Year's sales. Those plans were dropped when I heard the confirmation that the virus was spreading person-to-person: I borrowed a desk and emailed my editor and as many scientists as I could. My first report for *New Scientist* filed January 28th started with the words, "The new coronavirus may be about to go global."

That's how far things had already gone, and it wasn't speculation. Gabriel Leung at Hong Kong University is a leading expert in public health and a veteran of SARS. He and his team had also used travel data to calculate that dozens of infected people had long since traveled from Wuhan to China's bustling metropolises: Beijing, Shanghai, Chongqing, Guangzhou, Shenzhen.

On January 27th, he told a press conference that according to his mathematical models, without "substantial, draconian measures limiting population mobility"—even more restrictive than China had already imposed—epidemics outside China were inevitable. His model forecast 200,000 cases by the following week.

Three days earlier, Chinese scientists had published clinical details of the first 41 patients in the leading medical journal *The Lancet*. Chinese doctors complained that the information should have been shared with them earlier, as they started to encounter cases. But it clearly could not have been published while the official story was still that this was nothing like SARS.

"Clinical presentations greatly resemble SARS-CoV. The number of deaths is rising quickly," they wrote. "We are concerned that the novel coronavirus could have acquired the ability for efficient human transmission"—in other words, better than the clumsy SARS virus. Scientists are good at understatement, but that deserves a prize: the day after the paper appeared, there were officially 2,000 tested and confirmed cases across China and, we can now calculate, probably at least 8,000 milder ones.

The Chinese scientists were clear about what was needed to manage the epidemic: reliable, quick tests for the virus. They also noted the discovery in 2013, by the Wuhan Institute of Virology, of very similar viruses in bats that were already capable of infecting human airway cells.

"Because of the pandemic potential of 2019-nCoV," they warned, the virus would have to be carefully watched to see how its transmission and impact changed as it adapted to humans.

Everything was there. Efficient spread. Need for tests. Pandemic potential. At that point, countries around the world should have started intensively preparing for the virus to hit. Some did. Most did not.

Despite apparent openness, it seems China delayed reporting the illness, the virus, and especially the all-important person-to-person spread. Possibly, with memories of SARS still raw, authorities were afraid to frighten people with the news that it might be back. Darker stories have emerged that support this view.

On March 11th, Dr. Ai Fen, head of the emergency department at Wuhan Central Hospital, told the Chinese magazine *Renwu* (*People*) that on December 30th, 2019, the hospital lab had sent her a test result from one of the mystery pneumonia cases. It said "SARS coronavirus."

A PCR diagnostic test matches genes from an infection to genetic sequences from known disease germs. It's very possible such a test would have identified the then-unknown Covid-19 virus as SARS: many of their gene sequences are similar. In fact, the official committee of virologists tasked with naming the new virus decreed on March 2nd that the two are the same species.

The SARS virus they renamed SARS-CoV-1, CoV for coronavirus. The virus that causes Covid-19 officially became SARS-CoV-2, like the second installment in a movie franchise—*SARS 2: This Time It's Everywhere.*

Back in December, however, Dr. Ai didn't know any of that. She told *Renwu* that the diagnostic report made her break out in a cold sweat. SARS was a nightmare for China, officially infecting 5,327 people and killing 349, many of them doctors and nurses infected while caring for patients. The hospital sent part of the sample that tested positive for SARS off to Shanghai so the virus could be properly sequenced.

Ai took a picture of the report on her phone, with "SARS coronavirus" circled, and sent it to other doctors in Wuhan, including an ophthalmologist, Li Wenliang. He passed a warning on to colleagues about patients with pneumonia quarantined in the emergency department. The news spread fast: the hashtag "Wuhan SARS" started trending on Weibo, China's substitute for the banned Twitter. It was censored.

The hospital told Ai that night not to spread information about the pneumonia cases, so as not to cause panic and "damage stability." The hospital's disciplinary committee reprimanded her.

Staff were told not to exchange messages about the disease, Ai told *Renwu*, and astonishingly, not even to wear protective masks and gowns for fear of causing alarm. After all, there is no need for such protection with a virus that is not supposed to spread between people. Zhang Jixian at Hubei Provincial Hospital bought her staff protective clothing to wear under their normal white uniforms. They didn't get official protective gear until after January 20th when China admitted the virus was contagious.

The Japanese newspaper *Mainichi* echoes Ai's story. In late

January, it reported that at 1:30 AM on December 31st—the night of Ai's test result—eight doctors in a group chat discussing the threat of an epidemic that the result might pose were summoned by authorities and told to write self-critical essays about spreading rumors.

They did. The crackdown silenced doctors. And that day, researchers at the University of Toronto have discovered terms relating to Wuhan and pneumonia started being censored on the popular messaging and livestreaming platforms WeChat and YY. "If I had known what was to happen, I would not have cared about the reprimand. I would have fucking talked about it to whoever, wherever I could," Ai told *Renwu*, in a translation carried by *The Guardian*.

As the epidemic worsened, Li Wenliang was hailed as a whistleblower. On February 7th, he died of Covid-19. "I am not a whistleblower," Ai modestly told *Renwu*. "I am the one who provided the whistle."

The mayor of Wuhan ultimately had to resign and accept responsibility for missteps, although before he did, he blamed Beijing for controlling what he could say about the virus in public. Those controls don't seem entirely gone. Ai's March interview in *Renwu* reportedly keeps mysteriously disappearing from Chinese websites. It has been kept alive by Western coverage and Chinese netizens.

Meanwhile, the virus Ai Fen was told to shut up about has gone everywhere. On March 11th, WHO director-general Tedros Ghebreyesus declared it a pandemic.

\*   \*   \*

So that, as far as I can piece together at this time from a range of reports, is what happened. It is of course probable that more will come to light, and accounts could change. But now we can start to ask the crucial question: Could this all have been prevented? Could the Wuhan outbreak have been stopped from spawning a pandemic?

This is one of the first big outbreaks that has been analyzed as it happened, using modern technology for rapidly sequencing viruses from different patients and working out which virus is descended from which on the basis of small, shared mutations. And what stands out is that the first few sequences from patients in China, says Andrew Rambaut of the University of Edinburgh—who specializes in the evolution of emerging RNA viruses like this—were "genetically identical."

The longer a virus circulates in a species, the more it acquires random small changes in the gene sequence. If this virus had jumped several times from different animals, or circulated for long in people, there would have been more genetic variation in the early infections.

So, says Rambaut, "I would say that it was definitely a single jump, and it was probably not before early November," which matches the timing of the first known cases. It could have been a jump from one animal to one human, or it could have been a few animals, all the same kind with the same infection, to a few humans—we don't have details of the earliest cases, he says, so we can't tell.

But this does tell us there was no hidden epidemic happening over a larger area or a longer time, or there would be more variation. That means those first cases in Wuhan were all there were. In theory, if Wuhan had enacted stringent containment the moment they spotted that cluster—apparently in late December—then actively looked for other infections and contained them while there weren't many, they might have stopped the infection from spreading far. It would have been even better if they had spotted it earlier.

To answer the question of whether they could have stopped the virus from spreading entirely, we need to know how much action would have been needed and whether authorities would have agreed to the disruption it would have caused, knowing what they knew at the time.

Andy Tatem and Shengjie Lai and a team at Southampton University in England have measured how authorities might have done it. Cases in China increased exponentially, as diseases do when there is nothing to stop them, until the *cordon sanitaire* was thrown around Wuhan. After that, and as similar travel bans and social distancing orders were imposed across China, case numbers stopped rising.

The impact was stunning: China's epidemic actually peaked in mid-February, a turning point predicted by epidemiologists outside China on the basis of changes in reported case numbers as controls were imposed, and confirmed by a WHO delegation to China in late February. By late March, China was reporting no new cases. The problem was now everywhere else.

In a dizzying analysis using a mathematical epidemic model and the seven billion anonymous location records per day logged by China's Baidu cell phone network, Tatem's team quantified how people moved between China's 340 major cities as travel restrictions went into effect after January 23rd. They measured how that travel related to the data on the spread of the virus. Given that, they then worked out how the virus would have spread if travel was the same as Baidu had logged during the same weeks in previous, normal years, with no travel bans.

They calculated that with unimpeded travel, provinces outside Hubei would have had *125 times* more cases by the end of February. "China's vigorous, multi-faceted response is likely to have prevented a far worse situation, which would have accelerated spread globally," they wrote. There would have been much more of the virus in the world—what epidemiologists call amplification—if China had not wrestled its epidemic to a halt. That would have made matters worse for everyone.

But if Wuhan had imposed travel bans before those five million people left for the Lunar New Year holiday, could it have stopped the virus entirely? Tatem's team found that if China had imposed the same control measures a week before January 23rd, it would have prevented 67 percent of its epidemic.

And implementing the control measures from early January— when Wuhan knew enough about the infection to shut the seafood market—would have cut China's epidemic to only 5 percent as many infections. Such a small epidemic might well have been contained, especially if other countries were also alerted to watch for, test, and contain any infected people who crossed their borders.

"Technically we certainly could have attacked it effectively at that point and maybe contained it," says Tatem. "It's easy to say this with hindsight of course. There was so little we knew about the virus at that point. That would have made it hard to act rapidly."

Rambaut thinks more could have been done. "The authorities in Wuhan spotted the outbreak as an unusual cluster of pneumonia, but then spent weeks saying there was no evidence of human-to-human transmission when in fact it was doing exactly that." They knew enough to act, and they missed their window.

All that would have been needed, he says, would have been surveillance to spot the outbreak early, then intensive containment and contact tracing, to break all the chains of transmission before there were many cases.

In fact, those were tools China already had. In 2003, the SARS coronavirus initially spread out of control in China, and eventually in countries around the world, because doctors' initial warnings about the outbreak were stifled, at first simply by local bureaucratic inertia. To stop that from ever happening again, in 2004, China installed a Contagious Disease National Direct Reporting System in every hospital.

Doctors were required to enter the diagnosis into the system whenever they encountered certain key infectious diseases, including pneumonia of unknown origin, the *New York Times* reported on March 29th. A suspicious cluster would show up on a screen at the China CDC in Beijing, without anyone having to get past any reluctant bureaucrats.

If something worrying appeared, central officials could launch intensive case finding and containment efforts. In an online drill in July 2019, 8,200 health officials tracked and contained a simulated infection brought in by a traveler who had been registered on the system.

There was a compelling reason for doing this besides avoiding a rerun of SARS. Several strains of bird flu that can infect and kill people have emerged in China over the past 25 years—we'll be looking at those later. The one saving grace of these bird flu viruses so far is that they cannot transmit between people, although research has shown they can evolve that ability. If one became transmissible, it might be catastrophic. A cluster of cases suggesting the emergence of a transmissible strain would have to be contained with extreme urgency.

Reflecting that, doctors were instructed to enter any case of bird flu they encountered in the national direct reporting system within two hours of diagnosis. The frequency with which individual bird flu cases have been diagnosed across China over the past decade—to judge from ProMED—suggests the system has been working. Luckily, no worrying cluster has yet emerged.

Perhaps when tests showed that the unusual pneumonia cases in Wuhan in November and December 2019 were not a new kind of flu, health officials relaxed. According to leaked internal reports, in December 2019, doctors were told not to report such cases to the automated alert system, only to local health officials, who were reluctant to pass on bad news. They were also reluctant later, as the local Party Congress was held in Wuhan that January: case numbers did not rise while it was in session.

It was as if someone took the batteries out of the smoke alarm that sounded too many false alarms—so it missed a real fire. Word seems to have reached Beijing of the mystery pneumonia only on December 30th after doctors leaked reports online—the same day Dr. Ai saw the diagnosis of SARS. That could be why China alerted the WHO on December 31st.

After that, according to reports in the Chinese press cited by the *New York Times*, Wuhan officials downplayed the seriousness of the disease. They set a case definition that allowed doctors to report cases of pneumonia to the automated system only if the patient had had some connection to the now-closed wet market or to a known patient—a strange case definition for a virus that isn't supposed to spread person-to-person. In Wuhan, the virus was spreading freely, so increasingly, people who got it did not have the requisite connections to the market or known cases.

Thus, Wuhan's official case numbers stopped climbing. It's worth pointing out that this happened elsewhere too: later, some US states and European countries would test people with Covid-19 symptoms only if they had contact with China or a known case, even though the virus was already circulating elsewhere—including locally. As a result, they turned out later to have far more cases than they had realized.

Finally, Zhong Nanshan investigated and reported the real situation to authorities on January 19th. The next day, after claiming no new cases in weeks, Wuhan suddenly reported 157—and was facing a *cordon sanitaire*.

If Wuhan had used its automated system and alerted China CDC, could it have done enough, early enough, to have contained the disease? The system was designed to trigger a full-blown containment response. The cases in December should have been enough to do that.

## Could We Have Stopped This Whole Thing at the Start?

Would local officials have followed through with something that was not a drill? This raises a perennial dilemma of public health, as I heard from Sylvie Briand, head of infectious hazard management at the WHO, when we were discussing problems like this some months before she was plunged into the Covid-19 crisis.

Containing a new infectious disease before it has spread far nearly always means reacting before it seems like a big deal, she says. There may be only a few clinical cases, but you know several times that number are already infected and incubating the disease, especially if it is very contagious and spreads early in the course of infection. Covid-19 ticks both boxes. You have to contain such things early before they escalate.

This can be difficult, as at that point officials often see the threat as too trivial for such disruption, scoffing that more people die falling down stairs—forgetting that unlike infections, stair accidents don't then multiply exponentially. Yet people made such objections in the early days of Covid-19. And if you do manage to contain the disease, then nothing happens. Officials may wonder why they spent all that money fighting a threat that disappeared, even though that was the point. I still get letters when I write about some new disease, saying something like, "Oh, well, SARS was supposed to kill us all, and it never did, so why should we believe this?" Well, because with SARS, we eventually listened to the warnings and managed to contain it. We were also lucky.

Wait until the threat is obvious, though, and you are usually too late. "First they accuse you of overreacting," said Briand, echoing many frustrated public health experts I have heard over the years. "Then the epidemic suddenly explodes, and they say you didn't act fast enough."

This is especially the case when you cannot contain a virus just by quietly isolating the few people who have it and the few dozen people they contacted closely enough to pass it on. Officials might be okay with something that minor. But it might not be so easy.

Using a massive set of data on real social interactions in the UK, Matt Keeling at the University of Warwick and colleagues found that using the official UK definition of contact—being within two meters of someone for at least 15 minutes—you have to trace and quarantine 36 people per case of Covid-19 just to catch and isolate four of every five people that case infected. That's a lot.

And contact tracing might not be enough. As we have seen, the Chinese subsequently discovered that the key to stopping Covid-19 is using social distancing as well as containment. The variable that matters—and about the only piece of epidemiology jargon you really need to know to get all this—is R0, the basic reproductive number.

This is the number of people each infected person passes the virus on to, on average, at the start, when everyone is susceptible. And we all were susceptible to start with, as this was a virus no one had encountered before.

That value for Covid-19 was originally calculated at between 2 and 3, making it more communicable than most seasonal flu, although later calculations found it might sometimes be higher for occasional people who seem to spread it massively, called super-spreaders. Rosalind Eggo at the London School of Hygiene and Tropical Medicine and her team calculate that for a virus with a basic R0 like that, contact tracing and isolation alone only work if there is little or no transmission before the virus causes symptoms.

Otherwise, an infected person will have too many untraceable contacts, because the contact occurred before they knew they were sick. And even if you find those contacts, they will have had more time for their own infections to incubate and may already have spread it before you can quarantine them. Covid-19 spreads as much as a day or two before you get sick. The many cases with very mild symptoms, or none at all, also pose problems for containment.

It's as if a virus with a high enough R0 is just too slippery to pin down easily. So the answer is to reduce the number of people infected by each person with the virus. That's what mitigation does: with less person-to-person contact, fewer people catch the virus from any given case, so you have to quarantine fewer people to break transmission. If a virus has an R0 around 2.5, Eggo and team figure you need to reduce contacts by around 60 percent to get the R value down to 1, the level at which the epidemic stops growing.

So even if Wuhan had thrown itself into containment at the start, it may not have stopped the epidemic without social distancing as well. Epidemiologists would very possibly have known too little about the virus at that point to make a case for such drastic measures. Even much later, with far less excuse, some Western countries were slow to admit the need for such disruption.

"Social distancing is the magic ingredient in control," says epidemiologist David Fisman of the University of Toronto, another SARS veteran. "I have no reason to think they could reasonably have known that massive social distancing was needed in response to what at first appeared to be just a disease cluster in Wuhan."

That's the trouble with a new disease, he says. "We all learn a lot week by week, and are all making mistakes. That's the nature of the beast, I think."

Tatem agrees: "You only have to look back at ProMED to see lots of unexplained small outbreaks that lead nowhere," like those earlier reports of undiagnosed pneumonia from China. We can't throw a city into lockdown for every one of those. How do we distinguish things like that from the real threats, so that we risk mass disruption only—or at least mostly—for things that aren't going to fizzle out?

"We do need to get better at early detection and identification of those outliers that have the potential to cause major outbreaks," says Tatem. But of course, we can't even try to decide which outbreaks really have legs if we don't know the outbreaks are happening in the first place. That's where China's smoke alarm should have worked.

Zeng Guang, the chief epidemiologist at China CDC, is quoted as telling the Communist Party paper *Global Times* that local governments "only partially" based their decisions on what the scientists told them, instead favoring "social stability, the economy and whether people could happily enjoy Lunar New Year." You can't cause much disruption if that's your aim.

A preference for secrecy and stability won out over the scientists' epidemic models at the typical crisis point in public health: when you need strong action even though it doesn't look to observers— or politicians facing the year's biggest holiday—like much is wrong.

So, the big question: Could China have stopped the epidemic from becoming the pandemic? The epidemiology suggests that they might well have been able to slow it down, although it might have been hard to stop completely, even if the automated system had been allowed to do its job in December. Just the effort, though, would have had an incalculable impact.

It would have meant telling the world that a dangerous, contagious pneumonia had emerged in Wuhan. If, over at ProMED, Marjorie Pollack had been able to post that in December or even January 1, and the WHO had announced it, the world's virologists and epidemiologists would have run for their labs and models and started furiously posting results, as indeed they did a few weeks later once the news was out.

The world's developers of vaccines and drugs and diagnostic tests would have got to work. Other countries could have started testing people who had traveled to Wuhan earlier. As more cases appeared, China might have been able to impose the social distancing that would have made the difference, perhaps before five million people carried the virus out of Wuhan.

Those things happened anyway, but an earlier warning would have given everyone a few weeks' head start. We've all seen now what exponential looks like. A short time, at the right time, matters.

There is no question that when China finally did act, it was awesomely effective, if socially and economically painful. Dye's team found that, normally, 6.7 million people travel out of Wuhan in the month after New Year's. This year, there was almost no movement. That bought other cities, and the world, time to prepare.

Eventually, 136 Chinese cities also shut down their public transport and 220 banned mass gatherings. Dye's team found that cities that did those things sooner rather than later had a third fewer cases during the first week of their outbreak: curves were flattened, and the number of cases each person infected was slashed. Their models showed that the Wuhan travel ban alone or the shutdowns in other cities alone would not have reversed the climbing epidemic curve, but both together did—and cut the cases China would otherwise have had by 96 percent.

Wuhan required people to report their temperature daily, and in some cities that were not locked down, stores took people's temperatures before allowing them in. Anyone with a fever could go to a "fever clinic" for testing. People with cases too mild for hospitalization were isolated in repurposed stadiums and conference centers. Contacts of infected people were traced and quarantined.

An international team led by the WHO went to study China's response to the epidemic in late February. They reported that China had successfully managed to bend a steadily rising epidemic curve sharply downward. It stopped the virus spreading in the community in every province outside Hubei—most transmission was within families. It was by any standards an amazing achievement.

Bruce Aylward, the Canadian epidemiologist who led the WHO team, was jet-lagged enough to betray more than the usual slight hint of his Newfoundland accent when he briefed the press the day he flew back from Beijing. But he said he was convinced the decline in case numbers was real. Doctors had spoken of scheduling ordinary patients again. Lines outside fever clinics had disappeared. A big trial testing an existing antiviral drug against Covid-19 was having trouble finding participants.

China's initial delay may have let the virus get away. But its subsequent massive crackdown bought the world time, said Aylward.

If the spread of Covid-19 out of China was terrifyingly fast, don't even try to imagine what it would have been without the anchors China slammed on its own epidemic.

"We now know what works against this virus. We know what to do," Aylward said. He dismissed claims that only China could have imposed the containment and social distancing needed—the rest of the world could follow their model, adapting measures to their own conditions. He just wasn't sure the rest of the world "understands the need for speed."

Most of it did not. The virus had already gotten a head start in Italy, the UK, the US, and elsewhere by the time a serious response was launched. In late March, no Chinese province outside Hubei had officially reported more than 1,500 confirmed cases, but 15 US states had—and most Chinese provinces have more people.

Some places, however, contained the virus without the disruptive lockdowns needed in China and the West. Hong Kong, South Korea, Singapore, and Taiwan probably gave the world its best model for how to react, by imposing containment early enough and backing it with widespread testing for the virus. Their success suggests what might have happened in China if it had let its Contagious Disease National Reporting System trigger a massive containment effort in response to the first case cluster.

That second wave of countries also leveled with people. In an astonishing public statement, Prime Minister Lee Hsien Loong told Singaporeans on Facebook as early as February 8th that, despite a strong containment effort, the virus would probably spread in the community, and outlined the self-isolation measures that would be needed, "so we will be mentally prepared."

"Fear can make us…do things that make matters worse, like hoarding facemasks or food, or blaming particular groups for the outbreak," he said. On the other hand, he described students who, as Singapore imposed quarantines on exposed people, were already delivering food to shut-ins, and business federations, unions, and public transport "going the extra mile" to keep things running. "This is who we are," he declared. At a time when some countries seemed in denial about the virus, it was a moving performance. Public trust, say veterans at the WHO, is essential for responding to a crisis.

These countries also had experience with a similar disease. In 2015, South Korea had an outbreak of MERS, which they got under control using hospital infection control and quarantine. And Hong Kong, South Korea, Singapore, and Taiwan were all hit hard by SARS. They knew the need for speed.

Hong Kong traced and quarantined contacts, closed schools, cancelled big events, quarantined arrivals from affected countries, and encouraged working from home. At the end of March, it had only 715 confirmed cases—94 asymptomatic—and 4 deaths. The measures actually cut the transmission of flu at the same time by nearly half. As in other epidemics, ordinary people's behavior—masks and social distancing—made the difference.

At university lectures in Singapore in March, a maximum of 50 students were allowed, they sat at two-meter distances, and a photo was taken of who sat where in case contacts needed to be traced later. Public spaces were not closed, but anyone entering one had their temperature taken, increasing public confidence as much as catching cases.

## Could We Have Stopped This Whole Thing at the Start?

South Korea had companies making Covid-19 tests by early February. National labs double-checked the test results as people were tested, effectively doing the usual validation trials of a new test on the hoof to save time. The US Food and Drug Administration insisted that trials be done on US-made tests before they were used on the public, adding to an already disastrous delay in testing.

On top of that, South Korea had invented drive-through testing by late March. Positives were isolated and contacts quarantined; by April, case numbers were falling, without severe social distancing. The story was similar in Singapore and Taiwan. The difference was the early start that China had missed. Digital privacy experts have valid concerns about the expanded electronic surveillance involved. But the virus was contained.

You didn't need a history of tangling with coronaviruses to do the right thing, though. The small Italian town of Vò in Lombardy kept the virus under control by testing everyone and then imposing isolation and quarantine as needed. This should have been possible in far more countries in that first onslaught of contagion, yet many utterly failed.

If nothing else, these successful responses proved that containment, started early enough, worked on Covid-19. They confirmed that earlier action in China might have limited the epidemic. But mistakes were far from unique to China.

Wuhan had its Guinness-record potluck. But on March 7th, as the pandemic took hold in France—and we all knew the virus was contagious—more than 35,000 people dressed as Smurfs gathered in Landerneau, France. The next day, France banned gatherings over 1,000 people.

In late March, 70 University of Texas students were among hundreds who crowded onto beaches for the traditional spring break, despite warnings; 44 of the 70 later tested positive for Covid-19 and undoubtedly gave it to others. All these reactions seem to be simple, psychological denial: the refusal of people who have rarely been at much risk from infectious disease to believe they really needed to take a so-far largely invisible threat seriously.

Five million people left Wuhan before the lockdown. But even that painful lesson was not learned in time to avoid it elsewhere. More than six weeks later, Italian authorities locked down northern provinces that were the initial hotspots for the virus. The news leaked the evening before, and people fled, carrying the virus all over Italy. The whole country was shut down the next day.

In many countries, social distancing was partial or delayed, to the point where curves were barely flattened. Testing was delayed or restricted, endangering health care workers and patients and preventing containment. Even as the WHO stressed that containment worked with this virus, some countries abandoned it almost immediately, including Switzerland, where the WHO is located.

And ideology trumped public health in many places. A US administration focused on threats from foreigners rushed to close borders—after the virus had already arrived in the US and despite science and experience showing this does little to stop viruses.

Those things are still playing out as I write, so this will not be an analysis of what countries did, beyond the very early days, to respond to Covid-19. Those analyses will be needed. For now, we can say that few covered themselves in glory—and we don't yet know the long- or even medium-term outcome even for those countries that did well at delaying the first wave of the disease, as the virus still circulates and people remain susceptible. The accusations and political fallout will rage for a generation to come.

For now, we can ask whether more openness and earlier containment in China might have prevented the pandemic. This is not to point fingers or throw stones—most of us, in that respect, live in glass houses—but so that the next time this happens, wherever it happens, we might do better.

The answer seems to be that stopping Covid-19 entirely might have taken faster action than any government could have managed. But earlier action was possible, and that might have slowed the epidemic enough to make Covid-19 much less damaging and perhaps, just perhaps, kept it from reaching pandemic proportions.

According to the Chinese Communist Party's official newspaper, China's Supreme Court admitted as much on January 29th when they ruled that authorities in Wuhan were wrong to censure the eight doctors for their online chat about a SARS-like virus back in December. "The information would have pushed the public to take preventive measures more promptly, which could have been a fortunate thing given the current efforts needed to contain the virus." Xi's government has even turned Li Wenliang into a posthumous hero.

The first official case in Italy was detected on February 20th. Italian public health officials did the right things: isolating, contact tracing, locking down towns with the most cases. But it was too late: the virus was already too widespread, and hospitals were eventually overwhelmed. In fact, Italian epidemiologists later discovered that the first traceable case in the country fell ill on the first of January. At the time, no one suspected a thing.

If every country had known what China knew by early January, if it had sounded the alarm sooner and it told the WHO we had a problem, what might we all have done to stop the virus?

We will be looking at ways all of us might try to do better next time. Through pandemic planning. Through global viral surveillance and response when we find something worrying. Through a binding international agreement to monitor and control pathogens, this time with teeth. Through scaring the daylights out of ourselves by looking at what a worse pandemic might do.

First, let me explain why I'm so sure this will happen again. Let's look at where these viruses come from.

## CHAPTER 2

# What Are These Emerging Diseases, and Why Are They Emerging?

> "A new disease every day, and the old ones
> are coming back."
> —Loudon Wainwright III,
> "Hard Day on the Planet"

Ever since the HIV pandemic, people from health experts to screenwriters have been predicting what the next one would be. Various kinds of flu? A super-transmissible Ebola-with-wings? A souped-up version of the common cold? A bioweapon or a therapeutic virus gone wrong?

There were scares and near-misses with bird flu, mad cow disease, Ebola, SARS, and MERS, then the swine flu pandemic in 2009 that turned out relatively mild, although it still killed. Now there's Covid-19.

Why is this happening, and will it continue? And more to the point, what will hit us next? There are more viruses where Covid-19 came from.

We should start by saying what we mean by pandemic, and disease events generally. An outbreak is one or a few cases of an infectious disease that is unusual, so it gets noticed. An epidemic is a larger version of that: more related cases of a disease than usual spreading in a group of people. An epidemic can be a regular event, for example, the spread of flu through a city in the winter. Endemic means a disease that rumbles on all the time, like tuberculosis or gonorrhea.

A pandemic is when an epidemic goes global. Some health authorities impose other criteria, like that it has to be severe, or uncontrollable, or new, but those are not consistent or universal. In fact, there aren't fixed criteria for when an epidemic is big enough to count as a pandemic, except for flu—and even those have been changed lately.

We are sure there will be more pandemics, and probably sooner rather than later, as our population climbs and we increasingly crowd into cities, without routine surveillance of the world's scarier viruses or agreed global means for working together to contain them, and as intensive global trade and travel continue to transport the illnesses that emerge to people everywhere. The chaotic global response to Covid-19—just the fact that it got away in the first place—should make that situation obvious.

We can't precisely predict which disease will be next, or when, although anyone who studies infectious diseases could, and in fact did, tell you years ago coronaviruses were prime suspects. There's even an official short list of the diseases we're worried about but aren't ready for. To understand this predicament, though, we need to look back at some recent history.

In 1972, one of the world's then-leading experts on human infections, Macfarlane Burnet, co-wrote the fourth edition of a medical textbook called *Natural History of Infectious Disease*. And in it he wrote something quite astonishing. "The most likely forecast about the future of infectious diseases," he opined, "is that it will be very dull."

He must have enjoyed the shock value of that statement. Scientists always talk up their subject area, especially when trying to entice students into it. Burnet had just won a Nobel Prize for helping work out how our immune systems attack germs, but not us. He knew the value of studying infectious disease.

But the point of studying infectious disease is so you can beat it. And he figured it had been beaten. The comment was an in-joke, a somewhat smug whoop of victory. And also advice to a young doctor: specialize in something else. We've got this.

After all, in 1972, smallpox was all but eradicated. Most childhood diseases, even the super-infectious measles and the dreaded polio, could be entirely prevented with vaccines, and had been in richer countries. Formerly deadly bacterial foes—diphtheria, anthrax, TB, typhus, syphilis, gonorrhea—could be killed with antibiotics. Cheap, easily obtainable drugs stopped you from getting malaria. Infections still plagued poor countries, but surely development would fix that. In the 1970s, the Yale and Harvard medical schools downsized their infectious disease departments.

Burnet acknowledged, of course—because he knew colleagues would cavil if he didn't—that there was always a risk of "some wholly unexpected emergence of a new and dangerous infectious disease." But this struck him as unlikely. "Nothing of the sort," he assured the reader, "has marked the last fifty years."

How about the subsequent nearly 50 years then? Let's think. The deadly bacteria of Legionnaires' disease appeared four years after he wrote that. The US recognized the AIDS pandemic four years after that.

Then Lyme disease. SARS. MERS. Ebola. Marburg. Bird flu. Swine flu, another pandemic. Dengue. Chikungunya. Zika. Sin Nombre hantavirus. Nipah. Hendra. Lethal versions of normally harmless *E. coli* bacteria. Gonorrhea that resists all antibiotics. Ordinary urinary tract infections that resist all antibiotics. Extensively drug-resistant TB. West Nile. Mad cow disease, in cows and people. Oh yes, and a Covid-19 pandemic.

I wonder what Burnet would have made of the year 2020. Covid-19 is a lot of things, but it isn't dull.

What an eminent scientist thought in 1972 might seem like ancient history, but it matters to the Covid-19 saga. After infectious disease faded as a leading cause of death, and people everywhere started living to previously rare old age, the big killers—in rich and increasingly in poor countries—became conditions linked not to pathogens but to genes, environment, and lifestyle: cancer, heart attacks, strokes, Alzheimer's, traffic accidents, the complications of smoking and obesity. (There is very recent evidence that bacteria may be involved in big killers like Alzheimer's and heart attacks, but that's a different matter.)

Tackling those challenges didn't require the community-level public health that was historically designed for communicable disease, which involved quarantines and vaccination drives, not admonitions to eat more vegetables. The new killers certainly didn't require investment in new vaccines or antimicrobial drugs, pathogen surveillance, or local agencies that could monitor and contain epidemics. Nearly all of those capabilities atrophied, even in the richest countries.

Despite increasing alarm among researchers and global health experts about emerging infectious disease for almost three decades now, the mainstream attitude, especially in rich countries, has been complacency—perhaps because, as always in public health, problems tend to be invisible until it's too late. The old infections seemed to be gone, or problems only for the poor or marginalized. New infections seemed merely theoretical.

In response, medical industry shifted. Vaccines used to be made by government agencies as a public good, not for profit. The vaccine that eradicated smallpox, for example, was largely made by the Soviet Union and New York state. By the 1980s, vaccines were privatized, and in many cases, profits have been too low to encourage new investment. Most flu vaccines are still made using chicken eggs, a slow and sometimes problematic process from the 1940s.

Investment in public health decreased in many countries. In the US, there was a brief infusion of cash into preparedness for perceived bioterror threats after the anthrax mailings of 2001. But funding for the Public Health Emergency Preparedness agreement among state and federal agencies fell from nearly a billion dollars in 2002 to $675 million in 2019.

The minimal importance accorded public health was reflected in widespread cuts after the financial crisis of 2008. There has been a surge in hepatitis, Legionnaires', and diseases transmitted by sex or drinking water across the US, which public health experts attribute to health departments losing a fifth of their employees during that time. This is now hampering efforts to contain Covid-19.

In Europe, too, investment in public health plummeted after 2008. In 2019, a British think tank calculated that public health spending in England had fallen by £870 million just since 2014 and that this may have caused 130,000 deaths and a rise in chronic conditions, like diabetes, that incidentally also make you more likely to die from Covid-19.

The same thing happened with infectious disease monitoring and research in developing countries, where a network of labs largely left over from the colonial era disappeared, seen by former colonial powers as expensive anachronisms in the 1970s. Among those decommissioned was the British lab in Uganda that identified Zika and 30 other new viruses between 1930 and 1970. What if that lab had survived, increasingly with Ugandan scientists in charge, to recognize HIV in the 1970s? As we all know now, early action can make huge differences in a pandemic.

To appreciate the short-sightedness of this dismissal of infectious disease, let's look at some actual ancient history. In the ten millennia since we invented agriculture, infection has been by far the biggest killer of humanity, despite impressive competition from war and famine.

According to virologist Ab Osterhaus of the Research Center for Emerging Infections and Zoonoses in Hannover, in 1900, infectious disease caused fully half of all human deaths. Malaria alone is thought to have killed half of all humans who ever lived. (Those statistics don't contradict: the deaths involved different human populations over different times.)

In the 1800s, tuberculosis, known as TB, infected 70 to 90 percent of Europe's city dwellers and caused more than a third of all deaths, spawning a score of "consumptive" characters in Victorian novels. Yellow fever killed most of Napoleon's army in the Caribbean, so he unloaded the Louisiana Purchase on the US and abandoned the unhealthy New World. Many, if not most, children until very recently died before the age of five, nearly always of infection. In a few places, they still do.

Both illnesses and deaths by infectious disease in the industrialized world plummeted after 1950 and declined in many developing countries too. By 2004, infectious disease caused less than a quarter of all deaths worldwide, mostly in poor, tropical countries. In rich temperate countries, it was barely a few percent.

Many things contributed to this astounding decline. Besides drugs and vaccines, there were sanitation and hygiene. There was also vastly better nutrition, as chemical fertilizers and crop breeding boosted agricultural yields, and refrigeration and railways distributed fresh food, with the added bonus of banishing disease-ridden livestock, like milk cows with TB, from cities.

That's the big picture. There are countless little ones. I'm just old enough to have had measles at three, and I'm told I almost died of the common bacterial complications, saved by a large—and, I clearly recall, painful—midnight injection of what my mother said was penicillin. A few years later, my little brother got the new measles vaccine.

Mothers who listen to the lies of anti-vax campaigners today have never seen how measles, typhoid, and polio can carry off children. Afghan mothers have: in 2006, as aid agencies tried to remedy years of atrocious health care under the Taliban-led government, mothers and children waited for days outside clinics offering childhood vaccinations. They had seen the alternative.

So in the 1970s, infectious disease did seem to be on the run. When I was taking medical classes as a research student in the 1970s, my medical colleagues were given Burnet's message not to waste their time on germs. Curing cancer was the future: US president Richard Nixon declared war on it in 1971.

When Peter Piot, now head of the London School of Hygiene and Tropical Medicine, was a student in Belgium, his professors advised him not to specialize in infectious disease. Luckily, he ignored them, helped discover the Ebola virus in the Congo, and later led the global fight against HIV.

Because germs are not gone. As Jeff Goldblum stammers in *Jurassic Park*, life finds a way. Where there are billions of humans to parasitize, some parasite will find us. (Technically, pathogens are parasites, living off the work our bodies do to marshal the energy and tissues that maintain us.)

And the most insidious are the tiny viruses, most little more than a protein shell, maybe with a film of fat, enclosing a clutch of genes, made of either DNA, like our genes, or RNA, the mirror image of DNA we use to translate genes into protein. Viruses carry no energy-capturing or -processing equipment of their own, but use their few proteins to invade and hijack our cells, so they can use them to replicate and spread.

During the 20th century, we defeated most of the viruses we knew about, mostly with vaccines. However, we didn't realize there were a lot of viruses we didn't know about, which could jump to us from other animal hosts and cause havoc. The buzzword here is "spillover."

What Burnet didn't realize about the "wholly unexpected emergence of a new and dangerous infectious disease" was that the past 50 years, during which he thought no new diseases had appeared, were no guide to the next 50.

The first big shock was AIDS, recognized in the US when gay men started developing rare cancers and pneumonia because their immune systems were suppressed. In 1983, this was traced to the human immunodeficiency virus, HIV, which invades white blood cells of the immune system. By 1984, HIV was found to be widespread in heterosexual people as well in central and eastern Africa.

For a virus that works slowly and is relatively hard to catch—as we all know, it requires the mingling of body fluids—HIV went pandemic shockingly fast. Some 40 million people are now living with it worldwide, and it has killed 32 million since it was recognized.

HIV shows better than anything else why Burnet's victory whoop was premature. HIV is a chimpanzee virus that jumped to people around 1920, in southeastern Cameroon, probably when people ate chimp meat or got chimp blood in a cut. Researchers believe this sort of viral transfer happens frequently in people who interact closely with animals.

Most such viruses are too ill-adapted to people to settle in and cause an infection, and our immune systems clear them out quickly. A few would have successfully infected us—but when humans were nearly all subsistence farmers, living few and far between in small-ish villages, and rarely traveled, those viruses would have killed a few people, immunized any survivors, run out of victims, and died out in people.

HIV had probably been jumping into the occasional human and not getting much farther ever since its predecessors jumped from monkeys into chimps long ago. But around 1920, the group M strain of the virus hit the big time, when someone carrying it took a boat downstream from Cameroon to the regional boomtown of Leopoldville in the Belgian Congo—now Kinshasa, capital of the Democratic Republic of the Congo.

How do we know all this? In 2014, virologists led by Oliver Pybus at Oxford in Britain and Philippe Lemey at Leuven in Belgium studied some 800 HIV viruses from blood samples in old Congo medical records; the oldest was from Leopoldville in 1959. Their genetic sequences differed slightly, showing they had already been circulating and acquiring small mutations. Those mutations allowed the team to work out which virus was descended from which and how much time that took, and then construct a family tree. They were all descended from a common ancestor that infected someone around 1920.

Today, Kinshasa is the second largest French-speaking city in the world after Paris, and already in 1920, Leopoldville was no village. It was the capital of the brutal Belgian colonization of central Africa, with 15,000 inhabitants. With men pouring in from all over the region to find jobs, the sex trade was brisk. It was also brisk along the rail line to the copper, cobalt, and uranium mines in the southern Katanga region. Tens of thousands of men migrated from around Kinshasa to work in Katanga, and the sex trade followed. The team found the most genetic diversity in HIV samples from Katanga and Kinshasa, meaning the most infections were there.

## What Are These Emerging Diseases, and Why Are They Emerging?

There was another surge in virus diversity after Congo gained its independence in 1960. Initially, it was largely due to the re-use of needles, a good way to spread HIV. But then wars and upheaval after independence led to a steep increase in poverty. Jacques Pépin, of the University of Sherbrooke in Quebec, has calculated that the number of regular customers per female sex worker in Kinshasa shot up from a few, longterm regulars to up to 1000 different men a year, leading to a huge surge in infection. Haitians and other foreigners working in Congo left, some with HIV.

So group M HIV went global. It was simply in the right place at the right time—right, at least, for the virus.

AIDS raised some unsettling realizations, and in 1992, the US Institute of Medicine (IOM) issued a widely-read report on them. Human numbers were at an all-time high, as was global trade and travel—the word globalization had just become common. International disease surveillance was diminishing, just as infectious diseases could travel anywhere more easily than ever. "Profit and liability concerns" had cut incentives for companies to make drugs and vaccines for poor countries, the IOM wrote.

This all added up to a "danger of emerging infectious diseases and the potential for devastating epidemics," they concluded—like the one that had just emerged. Prejudice against the gay men who made up most of the early cases certainly slowed the response to AIDS unforgivably. But even if that had not been the case, it would have still been true that a previously unknown, horrendous virus suddenly arrived and spread and took the medical world completely by surprise. How many more of those were out there?

Yet the scientific and medical communities, the public, and

politicians all seemed complacent, not just about infectious disease in the US, but globally. "Complacency," the report warned, "can also constitute a major threat to health."

Just because we have suppressed some infectious disease, the IOM wrote, people seem to think we can readily suppress any of them—but old diseases can re-emerge, and new ones can emerge. The good news was that we could do something about this. "Anticipation and prevention of infectious diseases are possible, necessary and ultimately cost-effective."

How right they were. The cost of this Covid-19 pandemic runs into trillions of whatever currency you may name, and indeed beyond what can be accounted in money. In 2016, the US National Academy of Medicine, in a report trenchantly entitled, "The Neglected Dimension of Global Security: A Framework to Counter Infectious Disease Crises," calculated that dividing the expected cost of future pandemics up into cost per year amounted to $60 billion a year—a figure we might now regard as an underestimate. They figured you could prevent them for $4.5 billion per year.

Back in 1992, the writers of the emerging diseases report realized that, while all diseases are unique, some features of the AIDS story are typical. It is human ecology, more than anything else, that drives our illnesses. And economic globalization, changes in food production, and population growth were all profoundly changing our ecology.

The other important realization was that our infectious diseases mostly start in animals. The rinderpest virus—a major disease of cattle that was eradicated, after a long campaign, in 2011—evolved in us into the measles virus in the 11th or 12th century. Flu comes from ducks, smallpox from rodents, malaria from birds, mumps—we think—from pigs.

It's no accident those animals are mostly livestock or farm pests. We started living in large numbers and cheek by jowl with animals when we started depending on agriculture, around 10,000 BCE. With crops supplying rich, reliable food, our numbers exploded, and most of us settled near the fields, rather than continuing to wander as hunter-gatherers.

Viruses need hosts. To maintain itself in a human population, a virus needs a constant supply of fresh, non-immune humans, so it can move to a new victim before its current host either dies or develops immune reactions that finish off the virus. That requires a nearby, constantly renewed human population. Measles needs several hundred thousand people, the size of some medieval communities, to persist.

When we started living in groups that big, pathogens that had been exploiting herds of livestock and other hangers-on began exploiting our numbers. Now we are unprecedentedly numerous, and we are once again encountering a new source of viruses: the wild. HIV was a good example, but there are many others, not least Covid-19.

Peter Daszak heads the EcoHealth Alliance, a non-profit that conducts research aiming both to prevent pandemics and promote wildlife conservation. An Englishman in New York, he became captivated by wildlife diseases in 1995 after discovering a previously unknown pathogen causing diarrhea in a zoo's collection of giant hissing cockroaches. A natural showman, he once carried a pocketful of them into a TED Talk.

Until then, wildlife biologists hadn't been much interested in disease. It wasn't considered important to species survival. They reasoned that as a disease kills a species off, new victims become scarcer, so the disease fails to find new hosts and dies out long before the species does. After the pesticide DDT decimated birds worldwide, chemical pollutants got more attention.

Then, in 1997, a British lab found that a parasite can in fact drive a species to extinction, on two conditions: it parasitizes more than one species; and one of them tolerates the parasite and keeps it going, even if other host species disappears. North American gray squirrels have displaced native red squirrels in much of Europe partly because the grays tolerate the squirrel pox virus. The reds do not. In 2002, the Eurasian West Nile virus exploded across North America, causing occasionally fatal human infections but also slaughtering native birds, especially in the crow family, because they had no resistance. But sparrows, originally Eurasian, kept the virus going because they can carry it with no ill effects.

In 1998, Daszak was part of the team that found that just such an effect was allowing a previously unknown family of fungi, chytrids, to cause a massive worldwide die-off of amphibians that had driven some species extinct.

So wildlife biologists started learning about wildlife diseases. Eventually, it became clear that these diseases were also affecting humans. In 2008, Daszak and his colleagues calculated that of 335 novel pathogens that had emerged in people since 1940, 60 percent jumped to us from animals, and 72 percent of those, like Ebola and West Nile, came from wildlife.

The official term is zoonosis, from the Greek words for animal and disease. The team also found that the rate at which zoonoses were appearing was rising, as was the percentage of them that came from wildlife as opposed to domestic animals.

Again, the basic problem is our own growing population. When Burnet was writing, there were nearly four billion people in the world. Now there are twice that. More people demand more land and timber, and more jobs—and for some, that means catching wild creatures for ever more numerous city dwellers, for uses from pets to medicine. And especially, more people demand more food, so farmers scratch new farms out of forests and turn wild animals into new delicacies. The rest of us crowd into cities along with disease-carrying insects—and other humans.

Daszak and his colleagues mapped where the most new diseases had been observed and found "hotspots" in tropical and subtropical developing countries where economic development was creating concentrations of humans close to many species of wildlife.

This makes sense. The number of species of all kinds increases steadily as you get closer to the equator. There is just that much more, ultimately solar energy flowing through the system. More species mean more pathogens.

As species disappear under the onslaught of deforestation or other ecosystem destruction, they at least take their pathogens with them. But in degraded ecosystems, the remaining animals can also carry more pathogens than they might in healthier surroundings, because they are stressed or hungry, and germs take advantage.

Some biologists suspect a more insidious effect. When pathogens are hosted by several species, some hosts may limit the pathogen's numbers while others do not. When an environment is degraded, only one kind of host is often left—and those survivors tend to be the "weed" species that live fast, die young, and don't invest much energy in fighting pathogens. As a result, there might be a greater load of pathogens in the hosts left in a depleted ecosystem than in the original, diverse one.

There has been an increase in Ebola outbreaks since 1994, and researchers suspect they are associated with deforestation, which both displaces and stresses the bats that host Ebola and attracts more humans into bat country. The biggest-ever outbreak of Ebola raged through Guinea, Liberia, and Sierra Leone during 2014, killing at least 11,000 people. It started in the village of Meliandou in Guinea, where the original dense forest had been largely replaced by cacao, coffee, and other farms.

That left forest bats looking for a new home. Fabian Leendertz of the Robert Koch Institute in Berlin and a team of investigators visited Meliandou after the epidemic, and they found that the village children played in the huge, hollow stump of a rainforest tree near the village, the only remnant of the old forest. A colony of insect-eating bats lived in the stump, of a species that can carry Ebola. Two-year-old Emile Ouamono managed to contract Ebola, says Leendertz, although he doesn't know if the child was playing with a dead bat as has been reported: his family might have known, but like Emile, they died.

Most disease surveillance doesn't happen in these high-risk environments, though. It happens where the money and scientists are, in rich temperate countries, even though new diseases, like Covid-19, are far more likely to emerge in these hotspots. There are many hotspots in China, and also India and Indonesia, partly because of large human populations.

The EcoHealth Alliance says the answer is close surveillance of hotspots for early clusters of disease; research to identify new, potentially zoonotic pathogens in wildlife; and efforts to conserve that wildlife, so it will stay healthy and deep inside wilderness areas away from people.

In the next chapter, we'll look at how that failed with Covid-19. But for now, let's look at how a pandemic like Covid-19 might happen again if we fail to contain some of the other viruses experts think are especially threatening. Some of them are worse than the one we are fighting now.

In fact, disease experts all seem to agree on two things: another pandemic is coming, and no one can predict which pathogen will cause the next one. However, in 2016, the WHO and a panel of scientists decided that some pathogens do bear more watching than others. They made a "blueprint" for R&D (research and development) to equip humanity with vaccines, drugs, and diagnostic tests for the most worrying of these pathogens before they go rogue, and created a list of nine priority viruses for which they figured these should be developed fastest.

The list has already been updated several times, not least to accommodate Covid-19, which was unknown when the WHO and a panel of scientists made the original selection. To be fair, though, they didn't completely fail to predict this pandemic: the first list did include "coronaviruses." We knew the risk.

Most of the priority pathogens, however, are not previously unknown viruses lurking in wildlife in a hotspot. All but one were chosen because they already cause human disease, and they have been traveling and adapting, which is worrying—especially as we don't have remedies.

Their names at least are older and lovelier than Covid-19: Crimean-Congo hemorrhagic fever, Rift Valley fever, Lassa fever, Zika, Nipah, Ebola.

The one exception was dubbed Disease X, an instant hit with newspaper headline writers. It simply means a previously completely unknown pathogen we can't even guess at now, like the ones in wildlife Daszak and his colleagues warn about. It is on the list so research can be done into ways to respond to surprises, such as vaccine "platforms" that can quickly be adapted for a completely unexpected virus. We'll look at that later. Meanwhile, here are the viral Most Unwanted.

**Coronaviruses** get two entries on the latest list: one for Covid-19 and one for the two we already knew about, SARS and MERS. The Coalition for Epidemic Preparedness Innovations (CEPI), set up in 2017 in Oslo to fund vaccine R&D for potential pandemics, had five vaccines in the works for MERS, the only severe coronavirus circulating in humans before Covid-19 appeared. As of this writing, it has nine for Covid-19, all at very early stages.

Two more viruses on the WHO list belong to a family called Bunyaviruses. **Crimean-Congo hemorrhagic fever** virus lives in ticks across Asia, Africa, and southeastern Europe. It usually causes just a mild fever in people, but can also cause severe disease, killing up to 30 percent of cases, with, according to the European Centre for Disease Prevention and Control (ECDC), fever, dizziness, sensitivity to light, and "sharp mood swings" in which "the patient may become confused and aggressive."

There is an old Soviet vaccine of unknown efficacy used in Bulgaria, but it isn't widely approved, partly because it is made using mouse brain, which can cause problems. A European research project aims to find a better one—CEPI has two in very early trials. Meanwhile, the virus is invading new territory as global warming moves the ticks north: it appeared in western Europe in 2010 and has taken up residence in Spain.

**Rift Valley fever** is a Bunyavirus carried by mosquitoes mainly to cattle, but people can get it from mosquitoes or infected meat. It is found across Africa, but it spread to the Arabian Peninsula in 2000. It is often mild, but occasionally causes inflammation and bleeding in the liver, encephalitis—inflammation of the brain—and blindness. Half of severe cases die. Livestock can be vaccinated against it, which is promising.

**Lassa fever**, from another family, infects 500,000 people a year across West Africa, and again, most have mild or no symptoms. However, a few get severely ill, and 5,000 a year die. It is carried by the common multimammate rat—yes, that means it has more teats than other rats—so you would think it was unlikely to spread beyond the rat's habitat. But worryingly, it has occasionally shown that it is capable of some human-to-human spread.

Besides, better management of Lassa might help manage other

risky pathogens in the same region. In the West African Ebola outbreak in 2014, early cases were initially mistaken for Lassa, which helped the far more contagious Ebola spread. The WHO wants diagnostic tests to fix that. CEPI has six vaccines for Lassa in animal tests.

And there's a further concern with Lassa: it has relatives. In 2008, one that was previously entirely unknown killed a 36-year-old woman in Zambia. At the time, virologists warned how frighteningly little we know about viruses in Africa—where humans have lived longest and which therefore should, in theory, have more pathogens adapted to us than anywhere else. Five of the eight named entries on the WHO's priority list are originally African viruses.

One is **Zika**, a flavivirus, a family that already counts two notorious diseases, dengue and yellow fever. All three are carried by *Aedes* mosquitoes, named with the Greek word for "odious." One notorious *Aedes*, the aggressive tiger mosquito, is migrating outside the tropics with global warming and the global trade in used tires, which harbor puddles where it breeds.

That mosquito migration has already spread emerging disease. Chikungunya, a painful but usually non-lethal *Aedes*-borne virus from East Africa, had a mutation that adapted it to tiger mosquitoes in 2005 and began a series of explosive outbreaks around the Indian Ocean; it reached Italy in 2007 and the Americas in 2013.

Zika was discovered in monkeys in Uganda in 1947 and then spread in monkeys to Southeast Asia. Until 2006, there had only ever been 14 known human cases. What it did next was completely unpredicted.

In 2007, Zika caused a big outbreak on the island of Yap in Micronesia, followed in 2013 by French Polynesia and other Pacific islands. The virus's genes showed it came from Southeast Asia, either in an infected human or infected mosquitoes. The insects regularly hitch rides on airplanes, causing cases of "airport malaria" in countries that do not have malaria locally. Zika had been fairly mild the few times it was seen in people. But on Yap, some cases developed a potentially paralyzing nerve disorder, Guillain-Barré syndrome, an occasional complication of several different infections.

Then in 2015, it appeared in Brazil and rapidly spread across South America and into North America. This time, it was accompanied by severe birth defects in babies born to infected mothers, especially microcephaly, in which the head is abnormally small. Re-examined medical records revealed it had done the same in the Pacific, but no one had made the connection.

Just as they had done with HIV, Oliver Pybus and team at the University of Oxford sequenced the Brazilian viruses and discovered they came from Polynesia. In fact, they were all so alike that, as with Covid-19, Pybus concluded that they probably descended from just one introduction, perhaps only one infected person. That carrier's infection could have resulted from just one mosquito bite. One bite. By late November 2016, Zika had left 3,700 babies in the Americas with birth defects.

Pybus also found that Zika arrived in the Americas in 2013, two years before it was noticed—but as with any exponential epidemic, at first there were vanishingly few cases. It might have happened when Polynesian soccer fans traveled to watch the FIFA Confederations Cup games in Brazil in June 2013.

No one really knows why Zika suddenly started causing large

outbreaks. It did not undergo any obvious mutations that might explain it, Pybus found. It may be that everywhere else it has been, it mostly lived in monkeys but occasionally infected humans, and children got it early, had a very mild disease as children often do, and developed immunity. After one generation of this, no adult would be non-immune. On Yap, there were no monkeys, and no one was immune, so it hit everyone, and adults got a worse disease— especially if they were pregnant.

As for why Zika suddenly moved west, Pybus suspects it simply started getting more chances: flights from Polynesia to Brazil increased 50 percent between 2012 and 2014. To spread the virus across the Pacific Ocean, either infected mosquitoes have to make the air journey or an infected human does. For humans to carry it, they have to be bitten by mosquitoes at both ends of the trip: one to give the human the virus, and the second to get it and pass it on to another human.

Both kinds of events are made more likely by the fact that more people than ever before are now flying between countries in the Southern Hemisphere, including while it is peak mosquito season in both places. The UN's International Labour Organization reports that, since 2000, the number of people migrating for work within the global south, as opposed to migrating north, has grown from 60 to 82 million a year.

Eventually—and as epidemiologists predicted—Zika cases in the Americas subsided, as enough people had been infected to cause herd immunity. That happens when people who haven't had an infection and thus have no immunity become so rare that it is hard for a virus to reach someone new and susceptible before it dies out in its existing host.

Herd immunity wanes as new people, with no immunity, are born and accumulate, so epidemiologists expect Zika to return. They don't know when—it might be years. It might also migrate elsewhere. Two billion people on the planet live with tiger mosquitoes.

Experimental Zika vaccines had gotten as far as human safety trials by 2016. But they need to be given to people at risk of catching Zika to test if they work—and there is now too little Zika circulating to test them. Ironically, until it returns, we won't have a vaccine. This is a perennial problem with emerging diseases.

Meanwhile, managing Zika—and other mosquito-borne illnesses like malaria, chikungunya, and dengue—requires mosquito surveillance and control, and the 2015 epidemic revealed how little capability many health agencies now have to do that. Rich countries that don't have malaria, but used to have large antimalaria programs, abandoned them in the 1980s, and people and expertise went with them. In 2015, the US CDC found it had only 12 medical entomologists—scientists who can identify and manage mosquitoes that carry disease—to face the Zika invasion. The story goes that it pulled one retiree off a sailboat in the Caribbean to come back to work.

Then there's the one disease on the WHO priority list that is most likely to keep the scientists awake at night—other than flu, and we'll look at that later. Few people have heard of **Nipah virus**. Frankly, this one truly scares me.

Scientists who work on emerging diseases meet in Vienna every two years, and in 2016, I went along to talk to them about their research. As we chatted over good Viennese coffee, I conducted a straw poll: Which of these emerging diseases we're hearing about here scares you most? The hands-down winner was Nipah.

If you watched the last scene in the movie *Contagion* carefully, you know about Nipah. The virus is carried by Malaysian flying foxes, the world's biggest bats. Their species name is *vampyrus*, but they are fruit-eaters—with a five-foot, or 1.5-meter, wingspan.

In 1998, forest trees in Malaysia failed to fruit because of a drought caused by the cyclic climate variation El Niño, plus smoke from human-caused forest fires in Indonesia. That drove big, hungry fruit bats to farms in peninsular Malaysia, including a village called Sungai Nipah, where they ate farmers' fruit—and dropped half-eaten pieces into pigpens, as well as urine and feces.

Pigs being pigs, they ate the lot, and then developed a severe brain inflammation. People who cared for the sick pigs developed it too. The virus spread across Malaysia to Singapore, infecting 276 people in the two countries—of whom 106 died. A million pigs were culled in an effort to stop the spread of what was assumed to be a contagious pig disease. Then scientists figured out it was bats.

In 2001, Nipah turned up in Bangladesh and nearby India. It turned out fruit bats were drinking the sweet sap, toddy, that farmers were tapping from palm trees and contaminating it. The virus now causes a winter outbreak somewhere in the region every year, with death rates up to 75 percent. In 2018, it turned up 1,600 miles away from Bangladesh, across the Indian subcontinent in the southwestern state of Kerala.

The virus in Malaysia rarely spread person-to-person, but the Bangladesh virus does, although only to a few successive people before it dies out. The potential to spread more effectively may be there, though, infectious disease expert Daniel Lucey of Georgetown University explained to me. Most worryingly, he says, it sometimes causes pneumonia and appears to spread in coughed-out droplets. As we all know, diseases like that can cause trouble.

Kerala is widely considered to have the best public health in India—it has also done a good job of flattening its Covid-19 curve. It isolated and treated people who caught the Nipah virus and contained the outbreak, although 17 people died, including health care workers.

A very similar virus, Hendra, is carried by fruit bats and has spread to people via horses in Australia. Horses can now be vaccinated against it, which suggests human vaccines may be possible. In addition, The University of Queensland has developed a treatment. Antibodies are the proteins produced by your immune system that latch onto a particular pathogen and attract immune cells to destroy it. Various tricks enable immunologists to produce a line of cultured cells that all make the same antibody, called a "monoclonal."

The advantage is that production of a monoclonal antibody can be scaled up by making bigger cell cultures. You usually don't produce your own antibodies to a new pathogen until a week or more after you were infected, and if the initial disease is severe, that can be too late. An injection of monoclonal antibody that attacks the virus can help you fight it earlier. One that attacks Hendra virus—and should also attack Nipah—passed safety tests in people this year.

In 2018, the Queensland team sent the antibody to Kerala, but the outbreak was contained before they could use it. Kerala had only one case in 2019, a 23-year-old who recovered. The state is keeping the antibody in reserve, though we don't know yet how well it will work on Nipah. It will be good to find out: monoclonals are among the most promising treatments for viruses we might be able to brew up in a hurry. They are being investigated for Covid-19.

Once again, the perennial problem with developing remedies for emerging diseases is that they are just emerging. Virologists fear Nipah could evolve to cause epidemics, but for now, it strikes only small numbers of people randomly, so tests of therapies are hard to organize. CEPI has four vaccines in animal experiments or human safety trials, but to test whether a vaccine works, it has to be used in an outbreak.

No one knows where one will strike next, but it could turn up in new places. African fruit bats also carry the virus. In 2014, Daszak and his colleagues found that people in Cameroon living in areas undergoing deforestation—and who butchered fruit bats for food—had antibodies to Nipah, showing they had been infected.

Finally, there is **Ebola**. The epidemic that raged throughout 2014 in Liberia, Guinea, and Sierra Leone took the disease world by surprise. The virus, also carried by bats, was originally discovered in Congo and had previously caused mostly small, containable outbreaks there or nearby. No one expected it in West Africa, though it later emerged that the virus had been detected there earlier—but no one had paid much attention.

It started in December 2013 but wasn't recognized as Ebola

until March. It continued to spread as Guinea's government initially resisted reporting cases for fear of discouraging foreign investors. The WHO, reluctant to offend a member state and hamstrung by bureaucracy, also dragged its heels. When it hit the region's cities, the disease raged out of control, eventually infecting 28,616 people, 50 times more than any previous Ebola outbreak, and according to the most careful observations, killing 70 percent of them. By August 2014, when the WHO declared an emergency, the epidemic curve was exponentially headed for unthinkable heights.

Finally, the world—and the WHO, with a revamped effort led by Bruce Aylward, who later led its Covid-19 mission to China—responded and contained the epidemic with the same tools used for Covid-19: isolation, contact tracing, and quarantine. Also, as with Covid-19, changes in ordinary people's behavior were crucial. Friends stopped embracing, and families stopped touching virus-laden corpses at funerals.

As with Covid-19, there were no drugs or vaccines for Ebola. After the US anthrax scare in 2001, there was some funding to develop them, as Ebola was considered a potential bioweapon. The funding ended after a few years. Yet by early 2015, companies had taken the prototypes developed then and pulled off a world first: testing a vaccine in the middle of a raging epidemic.

One was nearly 100 percent effective, but it had taken a year to get it manufactured and approved so it could be deployed in Africa, by which time the epidemic was almost over and it could only be tested in a few places. That vaccine and another one were, however, deployed in the next Ebola outbreak in 2018 in the conflict-ridden east of the Democratic Republic of the Congo. That epidemic had almost been halted as of April 2020, almost unnoticed amid the rising Covid-19 pandemic. By then, the vaccine tested in 2014 had been given to nearly 300,000 contacts, and contacts of contacts, of people with Ebola, in a massive containment effort that was 97.5 percent effective at blocking transmission of the virus.

The WHO was heavily criticized for its slow response to the 2014 Ebola epidemic. But its role had always been to advise countries on medical treatment, set standards for medical products, and organize long-term efforts like vaccination campaigns. It was supposed to coordinate the response to international outbreaks too, but it was never designed to be a global emergency response agency. By 2016, it had undergone a major restructuring to become one. That has stood us in good stead with this pandemic.

So after dropping the ball on infectious disease in the 1970s, we have at least been talking about the renewed threat of them for years—as far back as that 1992 report that alerted the world to the growing threat of emerging disease. Yet it took a near-disaster with the world's largest-ever Ebola epidemic in 2014 to re-invent the WHO as the emergency response agency for epidemics we—astonishingly— never had up to that point. The Ebola emergency also launched CEPI and the WHO R&D roadmap, and its pathogens list.

Here's a rare, optimistic thought: think what Covid-19 might inspire us to do.

Yet it should be noted that, as a bat virus still unaccustomed to humans, Ebola spread relatively slowly in West Africa. Researchers have since found, to their horror, that as it spread, the virus adapted to people and may have gotten better at transmitting. That would make it much harder to contain.

## What Are These Emerging Diseases, and Why Are They Emerging?

Viral evolution is one of the larger unknowns humanity faces as it goes forward into a future where we are all more aware of how vulnerable we are to pandemic disease. We will look at that in more detail later, especially how that might affect some of the viruses on this list, as we turn toward the future.

First, we have enough to handle with the viruses we've got. Let's look at where Covid-19 came from.

## CHAPTER 3

# SARS, MERS — You Can't Say We Weren't Warned

> "Experience is that marvelous thing that enables you to recognize a mistake when you make it again."
>
> —Franklin Jones, 20th-century US journalist, quoted in 2006 by epidemiologist Zhong Nanshan in relation to SARS

"Have you heard of an epidemic in Guangzhou? An acquaintance of mine from a teacher's chat room lives there and reports that the hospitals there have been closed and people are dying." Stephen Cunnion, an infectious disease expert and former head of preventive medicine for the US Navy, received that email from a friend on February 10, 2003. He couldn't find any further information, so he passed the email on to ProMED.

The same day, ProMED got a notice from the Hong Kong

health department, warning travelers of a pneumonia outbreak in Guangdong, the southeastern Chinese province of 100 million people next to Hong Kong. Guangzhou is its capital and biggest city. ProMED posted both messages. The next day, the WHO asked China about it. The health ministry in Beijing replied that the province did have a pneumonia outbreak that had started the previous November. There had been 305 cases and five deaths.

That was the first the world heard about what was eventually called severe acute respiratory syndrome, or SARS. In the first half of 2003, it traveled to 29 countries and territories, infected 8,096 people, and killed 774, many of them health care workers. Then it was stamped out.

SARS comes up repeatedly in any discussion of Covid-19, as it was in many ways a forerunner. The Covid-19 coronavirus has been officially declared the same species as the SARS virus, named literally SARS-CoV-2. It spreads more readily and is less lethal, but otherwise, it is much the same. And looking back at reporting I did for *New Scientist* about SARS, I'm astonished at how little else seems to have changed.

To understand what unleashed Covid-19 and what needs to happen to prevent the next pandemic, we need to understand SARS. It was, after all, a very clear warning of what is happening now. And then we got two more warnings. And we still did very little.

The WHO already had an inkling something was happening: in late 2002, a Canadian government system that monitors global press reports for mentions of disease picked up reports of pneumonia in China. But the rules in force at the time meant WHO couldn't inquire further about information it had not received officially from a government. It couldn't ask China for details until February 11th, after the official Hong Kong warning.

The same day, Guangzhou made its first public statement about the outbreak, which had already led to local panic buying of herbal remedies and vinegar, a traditional disinfectant. The health department said it was caused by a common bacterial infection, mycoplasma, and was under control. ProMED called this "speculative" and posted a press report that many of the people hospitalized were doctors and nurses.

On February 18th, the China Center for Disease Control and Prevention (China CDC) said the outbreak was chlamydia, another bacterial infection, and under control. Again, ProMED was politely dubious. Both of the bacteria mentioned so far were good news in a way because they could be treated with antibiotics. But both could also be co-infections alongside a primary infection with a virus, for most of which there are no drug treatments. On the 20th, ProMED posted a press report by a foreign-owned news service that quoted a doctor in Guangdong—who did not wish to be identified—saying a virus could not be ruled out.

There was little the WHO could do. The rules at that time were that it could not even tell the world about an outbreak unless the country that had it gave permission. And it could not officially get information about an outbreak from any source but that country's government, leaving it unable to act on information it got anywhere else.

Then our increasingly interconnected world took things up a notch by moving the outbreak across a border. On February 22nd, Liu Jianlun, a doctor who had been treating the pneumonia in Guangdong, developed symptoms while staying on the ninth floor of the Metropole Hotel in Hong Kong, where he was attending a wedding. He warned hospital staff to isolate him, knowing what he had, but they had had no official warning of the danger and took insufficient precautions. Some were infected. Liu died ten days later.

Meanwhile, seven other people who stayed on the same floor of the Metropole developed pneumonia—and carried the virus to other hospitals in Hong Kong and to three other countries: Liu was a super-spreader. No one ever figured out exactly how they were infected. SARS traveled in coughed-out droplets like Covid-19 and persisted the same way on surfaces. Some suspect the ninth-floor elevator buttons.

Johnny Chen, an American businessman on the ninth floor, developed pneumonia a few days later in Hanoi. His carers also fell ill. Carlo Urbani, a 46-year-old Italian infectious disease expert working for the WHO in Hanoi, realized this was a new disease, warned the WHO, and started infection control measures at the hospital. He later developed SARS and died. There is a memorial plaque to him at WHO headquarters in Geneva. The virus cultured from his lungs, known as the Urbani strain, is still used in virological research.

Other people from the ninth floor traveled to Canada and Singapore: those countries eventually had some 250 deaths each. Vietnam had 63. China had 349, Hong Kong, 299.

SARS had a 10 percent death rate, higher than any estimates for Covid-19. It, too, hit the elderly harder, but scaled up: half the people over 60 who got it died. Both viruses seem to kill the same way: a runaway reaction of the immune system called a cytokine storm, named after the signaling chemicals the body uses to turn on the immune reaction called inflammation. Normally, inflammation is the way our bodies fight invaders—but in some people, some pathogens can unleash too much of it.

Now that the outbreak was outside China, the WHO could act at the behest of the other affected countries. On March 12th, it issued an alert, warning countries and airlines to watch for cases, and citing the outbreaks in Hong Kong and Vietnam—and Guangdong, even though authorities there still officially attributed it to chlamydia. The WHO noted that investigation of the cause in Guangdong "is ongoing." Soon there were cases in Taiwan, Singapore, Thailand, and across the world in Canada. But China still reported only 305 cases—the same number as a month earlier.

At the same time, it asked the WHO for technical assistance. A team arrived in Beijing on March 23rd. Immediately, China's case numbers jumped to 792, a suggestion that authorities were being more open now with other experts in the room. But the team was kept in Beijing and not allowed to visit the epicenter in Guangdong until April 2nd.

At that point, things came to a head. The principle of state sovereignty was enshrined in the WHO's founding treaty when it was set up, along with the UN itself, in the wake of World War II, and ruled in disease management as in everything else. The International Health Regulations, a 1969 treaty with antecedents going back to the 19th century, prohibited the WHO from doing much without a member state's explicit permission.

But as fears about emerging diseases rose after AIDS and the 1992 Institute of Medicine report, a new idea was taking hold: global health security. The idea was that in a highly interconnected world, diseases could go global fast, so sometimes, for the greater good, an international agency should have the right to interfere in a sovereign state to ensure threatening outbreaks are contained. Implicit in this was the belief—born of repeated experience the world over—that what governments do in their own interests might not be in the interests of the world as a whole, especially when it comes to disease.

In 2003, the director-general of the WHO, Gro Harlem Brundtland, a doctor and the former prime minister of Norway, decided to interfere. Brundtland is still revered by many at the WHO, partly for how she handled SARS. She seemed to have a taste for combat. She once jabbed me in the chest, hard, and then walked away when I asked her at a press reception for her response to some criticism (all she heard was that I had repeated the criticism). I asked one of her aides if she was always like that. "Oh yes," he said.

From the beginning, the SARS virus had demonstrated relatively limited spread in droplets, and all cases could be traced to other cases, meaning containment could work. But on March 30th, more than 200 cases suddenly appeared at the Amoy Gardens apartment complex in Hong Kong. David Heymann, at the time head of infectious diseases at the WHO, told me there were fears it had become airborne, which would make it much harder to control. So on April 2, 2003, Brundtland advised the world to cancel all but essential travel to Hong Kong and Guangdong.

Fears of airborne spread subsided when the Amoy Gardens outbreak was traced to faulty plumbing. But the WHO issued further travel advisories for Beijing and Toronto in late April. None were lifted until each city had largely contained its epidemic, after a week for Toronto but around two months for the other cities.

It was unheard of for the WHO to issue advice with such direct financial implications for countries without their blessing. The advisories meant considerable lost business: Toronto estimates it lost $265 million. It sent delegations to complain to WHO headquarters in Geneva.

China's health minister reacted more strongly to the travel advisory, publicly urging people the next day to visit Guangdong. Over the next few weeks, authorities limited visits within China by WHO officials and seemed to be understating case numbers. Brundtland was openly critical.

Then on April 9th, Jiang Yanyong, a retired surgeon at a Beijing hospital, told Beijing TV stations that claims the epidemic was under control were nonsense, saying there were more than five times the official number of cases just in Beijing. Western media reported it, and Chinese citizens spread the news on cell phone networks.

The next day, Zhong Nanshan, then the head of the Respiratory Disease Research Institute in Guangdong, told the press: "The origin of this disease is still not clear, so how can you say it has been controlled?" He suspected a virus and later wrote that, in fact, a Chinese lab had identified the coronavirus as early as February 26th but kept it quiet, as official sources blamed bacteria.

Then, on April 14th, a Canadian lab sequenced the SARS virus from a patient whose infection could be traced back to China. Further claims that bacteria were responsible were no longer credible, and the WHO confirmed the virus as the cause of the pneumonia.

Yanzhong Huang of Seton Hall University, who has researched the period, writes that on April 17th, the top committee of the Chinese Communist Party called for a change in policy, and China's president, Hu Jintao, instructed officials to stop withholding information about the epidemic. Two days after that, officials admitted that Beijing had 346 SARS cases. The reported number had been 37.

Outside China, the world had swung into action. Heymann set up daily conference calls between doctors, epidemiologists, and virologists around the world, to compare notes on the best treatment, speed up the development of tests to diagnose the virus, and work out how it spread. The worldwide conference call became a durable model, Heymann told me in March this year: the same groups were convened for Covid-19, this time online.

Then as now, what was crucial was old-fashioned epidemiology. Cases were isolated. Their contacts were traced and quarantined. Workers in affected hospitals self-isolated if they developed a fever. Hong Kong barred contacts from leaving the territory, and police went after quarantine breakers. Singapore monitored quarantine with then-novel internet cameras. Toronto eased up on quarantine too soon and almost lost control, but got it back.

However, at the time, senior health experts told me that while this containment might slow the epidemic, the virus would inevitably get into cities that simply did not have the resources or the social order required to contain it. SARS was here to stay. On April 26th, I wrote in *New Scientist* that rich countries, in pure self-interest, must ensure that any eventual vaccine went to both rich and poor—because "sooner or later, SARS is coming to a person near you."

But my sources were wrong, even though their fears were entirely reasonable. Sometimes, luck takes a hand: the virus never reached Kinshasa or Calcutta. On July 5th, there had been no new cases for three weeks, and the WHO declared SARS "contained."

What had also happened was that after abandoning its efforts to conceal SARS, China launched mass mobilization to contain the virus, confining Beijing students to dorms and spending more than $1 billion refurbishing hospitals and finding and isolating cases. Sound familiar? The abrupt shift from downplaying the outbreak to a no-holds-barred response was eerily similar to what happened with Covid-19. Then as now, it worked. And it would have been a lot easier—and lives around the world might have been saved—if it had happened earlier.

Since then, more of the underlying story has come out. Huang wrote in 2004 that Guangdong health officials initially recognized the new disease as a virus and alerted authorities—but the law made any infectious disease outbreak a state secret until the health ministry announced it, so they couldn't tell anyone else. Then there were bureaucratic delays at the ministry, some because of the Lunar New Year holiday, until February 11th.

Then the news blackout resumed during the National People's Congress in March, just as it did during the Party Congress in Wuhan with Covid-19. Lower-level officials soft-pedaled reports to senior officials, Huang found, for fear of looking bad. Again, there are parallels, he says, with Covid-19.

Having faced a harrowing near-miss on SARS, you'd think we'd be doing better by now. And in fact, in the aftermath, China installed the automated alert system we talked about in Chapter 1, meant to allow doctors to alert central authorities to certain diagnoses, notably undiagnosed pneumonia, bypassing bureaucrats to make sure the blockages that delayed reporting SARS didn't happen again. Yet it was sidelined when Covid-19 emerged by the same bureaucratic culture of suppressing bad news.

There were a few quickly contained outbreaks of the virus the year after SARS was beaten, including escapes from virology labs and some outbreaks described as having been from a "wild source." Both types were troubling.

Labs remain a worrying source of dangerous viruses, although scientists and regulators tightened up precautions after the SARS incidents. In April 2020, some alleged that the Covid-19 virus might have escaped from the Wuhan Institute of Virology, where China's first highest-level containment lab, the kind used to study the most dangerous pathogens, opened in 2015. No evidence for this was offered. But there are concerns with all these labs that despite the showers and filters and hazmat suits, a scientist might be infected with a virus that sometimes causes few symptoms, and then carry it outside.

And the possibility remained that the SARS virus—or something very like it—still lurked in the wildlife it had presumably come from in the first place. That in fact was discovered by the virologists at the Wuhan Institute, initially in 2005 and definitively in 2017, and they warned the world about it. We'll discuss that later.

But although some SARS virus remained in a few lab freezers and wildlife, it was clear that SARS was gone from people: a triumph, if a close shave. What was crucial in controlling SARS, says Heymann, was that unlike Covid-19, the virus could not be spread by oral or nasal droplets until late in the infection, well after symptoms started, because it did not build up in the nose and throat until then. Thus, if you isolated every exposed person with a fever, you had contained it. With Covid-19, people with symptoms have already been spreading virus around for a day or two. Viruses that spread before they cause symptoms are very hard to contain: look at HIV.

Because SARS didn't spread as readily as Covid-19, no severe social distancing was needed to slow its spread enough to cut the number of contacts that had to be quarantined and make containment possible. There were also no asymptomatic cases. So SARS never spread far in the community.

And it never got into any big, chaotic cities in poor countries that could not have contained it. That did not happen this time: the number of flights from China to such places has increased tenfold or more since then. That is partly due to increased travel globally and the vastly increased prosperity of many Chinese. It is also partly due to the Belt and Road Initiative, China's massive program for investment and infrastructure across Eurasia and Africa.

Simply put, in 2003, we dodged a bullet: ProMED, WHO leadership, global collaboration among experts, shoe-leather epidemiology, and eventually, massive Chinese action—led finally by its doctors and scientists—eradicated SARS. They were helped by the fact that the virus was simply clumsier at spreading among humans than Covid-19.

What stands out now is the speed and efficiency with which, in retrospect, the world acted. The virus didn't get opportunities to establish itself in countries outside China that delayed taking action to contain it, as happened with Covid-19. As the virus arrived, there was no disputing the need for containment or talk of relying on herd immunity. Because of this swift action, SARS never circulated widely enough to be called a pandemic.

Maybe the virus's high death rate scared everyone into line. Maybe its inability to spread before symptoms started, and absence of many mild cases, just made it easier and less disruptive to follow the epidemiologists' advice. And there was more public trust in experts 17 years ago.

But did we learn our lesson with SARS and apply that to its sibling, Covid-19? In its World Health Report in late 2003, the WHO listed the five top lessons from the epidemic.

Lesson five was that health systems should protect health care workers, who made up between one-third and two-thirds of

SARS cases in heavily affected countries. Most nurses were (and are) women, and among health care staff, women were 2.7 times more likely than men to get SARS, whereas outside of hospitals, it infected both equally. Yet even now in some rich countries, doctors and nurses are dying and must face Covid-19 with insufficient masks, gloves, and gowns. Lesson not learned.

Lesson four: "the world's scientists, clinicians and public health experts are willing to set aside academic competition and work together for the public health good when the situation so requires."

To say that has happened again with Covid-19 would be a massive understatement: the outpouring of scientific and medical collaboration has been astonishing, as has the sheer quantity of research, posted almost before the ink is dry on the data, on preprint servers like bioRxiv or medRxiv. This also means it's posted before the usual expert reviewers have formally decided it's okay, which can call for caution—but in many cases, other scientists have piled on and reviewed this research anyway.

"I'm constantly amazed at how easily all the technical people communicate," said Bruce Aylward of the WHO, after returning from investigating Covid-19 in China in February. For me covering the story, it's been amazing watching a global scientific community deal 24/7 for months with a truly global crisis. And with both viruses, it was doctors and scientists, in China and elsewhere, who told the world how serious things were. I'm not sure we had to learn this lesson; we already knew.

Lesson three was that travel restrictions can help, the WHO claimed, despite conceding that temperature checks in airports caught only two cases of SARS. This is tricky. Modeling by several research groups has suggested closing borders doesn't actually achieve much. With Covid-19, the WHO advised against it; closed borders certainly impeded the response to Ebola in 2014. On the other hand, countries the world over have famously closed their borders as an extension of their lockdowns to slow the spread of Covid-19, and travel restrictions were crucial in China. Let's say lesson learned.

Lesson two was that global alerts work. After the WHO issued its alert on SARS in March 2003, affected countries redoubled efforts and got the epidemic under control, and others kept imported cases from spreading. The International Health Regulations Treaty was massively revised in 2005 because of SARS and enshrines this by having WHO declare a Public Health Emergency of International Concern when an extraordinary threat looms. With Covid-19, they made that declaration on January 30th. Lesson learned.

Lesson one is worth quoting in full:

The first and most compelling lesson concerns the need to report, promptly and openly, cases of any disease with the potential for international spread. Attempts to conceal cases of an infectious disease, for fear of social and economic consequences, must be recognized as a short-term stop-gap measure that carries a very high price: the potential for high levels of human suffering and death, loss of credibility in the eyes of the international community, escalating negative domestic economic impact, damage to the health and economies of neighbouring countries, and a very

79

real risk that outbreaks within the country's own territory will spiral out of control.... Strengthening of systems for outbreak alert and response [is] the only rational way to defend public health security against not only SARS but also against all future infectious disease threats.

And the WHO meant that to apply to all countries. Even though the burden of sounding the alarm fell to China in 2003 and again in 2020, the whole world needs this lesson. China certainly was more open about many things from the start on Covid-19 than it was about SARS in 2003—except for the crucial detail of it being contagious. Lesson not learned.

So that's two learned, two not learned, and one not needed. Clearly, the lessons unlearned from SARS have been deadly.

Case in point: I wondered in *New Scientist* in April 2003 whether, if Hong Kong hospital staff had known more about this new pneumonia in February, they might have used better infection control and not let it spread further. Or if China had acted sooner, whether it could have limited SARS to Guangdong. We're wondering those kinds of things now about Covid-19.

If nothing else, though, you would think we would have some coronavirus drugs and vaccines by now, just in case SARS or something like it ever came back, which, of course, it now has. But China's bureaucracy is not the only system failure in this saga of global disease mismanagement. Western capitalism has its glitches as well.

The silver cloud of eradicating SARS had a dark lining. Vaccine labs and antiviral drug developers set to work as soon as the SARS virus was identified, and their findings are being dusted off now to fight Covid-19. But, those experts say, funding to continue the research dried up after 2005, precisely because SARS had been eradicated—so our knowledge and tools are nowhere near what they could have been if that work had continued.

With no virus circulating, it is, if nothing else, hard to test whether a drug or vaccine works, because normally you do that by treating infected people or by vaccinating people and watching to see if they get infected. You can look for alternative measures of success or failure, such as durable immune reactions to a vaccine in humans or convincing results curing (with drugs) or protecting (with vaccines) animals exposed to the virus experimentally, in a high-containment lab.

But no one bothered doing this, says David Heymann, because with no SARS virus circulating, there was no market for any SARS drugs or vaccines they developed. Only big pharmaceutical firms have the know-how and money to get a drug or vaccine through the big, complex safety and effectiveness trials they rightly need before governments approve their use. Without a market, the companies cannot invest in these expensive trials because they won't make their R&D investment back by selling the eventual products.

At one time, as we saw with smallpox vaccines, some pharmaceutical companies were state-owned and could undertake work for the public good. But since the 1980s, those have disappeared, and pharmaceutical development is all done by private companies that are required to turn a profit. It's not because they're mean, it's because we decided as a society to do it that way, influenced by ideas that as much as possible should be done by the market rather than government. With SARS gone, it was too big a financial risk for a private company to invest in SARS drugs and vaccines. There was no guarantee they would ever be needed.

The same market failure stifles R&D for other vital medicines that for various reasons cannot be sold in large enough quantities or at high enough prices to make the R&D investment back, most worryingly new antibiotics. Mechanisms for getting around this problem, by rewarding drug developers in ways not tied to selling the product, have been discussed, but little has been attempted on a commercial scale.

This all makes developing products for any emerging disease difficult or impossible for the private companies that do much of our drug and vaccine development. However, public good is making a comeback. Over the past decade, public-private partnerships have emerged to develop drugs and vaccines for diseases mostly found in poor countries, funded by the Bill and Melinda Gates Foundation and others. CEPI, launched just in time for this crisis, organizes such funding for vaccine R&D for emerging diseases, and now for Covid-19.

But there was another reason no real effort was made after SARS to develop remedies for coronaviruses: some virologists decided SARS was never coming back. They made two mistakes. Rolf Hilgenfeld, of the University of Lübeck in Germany, was working on a promising approach to anti-SARS drugs when funding for that research dried up in 2006. He says one mistake was based on a major genetic difference between SARS and other coronaviruses: in SARS, a swath of 29 nucleotides found in other coronaviruses was simply missing from one gene.

Such "deletions" are certainly not unknown in viruses like these, which keep their genes as RNA rather than DNA: RNA viruses tend to be less genetically stable. And the gene's function was then unknown. Yet, some argued this major change was what enabled SARS to suddenly spread in humans. And, they reasoned, exactly the same dramatic mutation was unlikely to happen again. Ergo, SARS was not coming back.

Other virologists disagreed. "I certainly never said that," says Ab Osterhaus, a leading virologist, whose lab did experiments in 2003 that clinched the proof that the SARS virus caused the disease. Those who did say it were at least right about the deletion not coming back. The Covid-19 virus does not have it. But Covid-19 spreads even better in humans than SARS did, so clearly the mutation didn't play the role they imagined.

But there was also a second mistake: thinking SARS was gone in wildlife, too. In 2005, the repeated failure to find the virus in civets led some researchers to conclude the virus was gone in nature and had disappeared as a threat.

People who fell ill with SARS were initially linked to a wildlife market, as were some of the first people with Covid-19. The SARS virus was found in Guangdong markets in cages that had been occupied by masked palm civets, a member of a family of mammals related to cats, which in China are bred on farms and sold as game meat. TRAFFIC, an environmental group headquartered in Cambridge, England, that monitors trade in endangered wildlife, says some 10,000 civets were destroyed in markets in China in 2003 in an effort to stamp out the virus.

Tragically for those civets, virologists now think they had little to do with it. The virus came from bats. A few civets and another mammal, the raccoon dog, in Guangdong markets were found to have SARS or to have been infected with it in the past, and attention focused on civets because far more of them were sold. But, said virologists reviewing the research in 2007, SARS was never found in civets anywhere else, wild or farmed, and the evidence suggests the animals were infected at the market the same as humans were. Nonetheless, the story that civets were the "intermediate species" that spread the virus to humans became widespread. Similar tales are now being told about Covid-19 and pangolins.

In 2005, though, Chinese scientists were already warning that the virus could well lurk in other species. Also that year, virologists at the Wuhan Institute of Virology reported coronaviruses that were very similar to SARS in bats, which were also sold at markets. "If no action is taken to control wildlife markets," Zhong Nanshan warned in 2006, the SARS virus might once again "develop into an epidemic strain." Nevertheless, companies and government research funding agencies went with the more welcome assessment that SARS was gone.

And yet, between the disappearance of SARS from humans and the arrival of Covid-19, we got another warning about these coronaviruses: MERS.

In June 2012, Ali Zaki, an Egyptian virologist working at a hospital in Jeddah, Saudi Arabia, couldn't identify what had killed a 60-year-old man with pneumonia. The only positive test was a generic one for coronaviruses. But SARS was gone, and the other known coronaviruses in humans cause common colds. If any virologist could identify an unknown virus quickly, Zaki thought, it's Ron Fouchier in Rotterdam. He sent him samples.

Normally, scientists can't afford to take time identifying odd viruses that crop up—"stamp collecting" they dismissively call it. You rarely get a publication out of it, and researchers' jobs depend on getting research grants, which are given for research that does result in publications.

But one good response to SARS was a program in the EU that funded researchers to do the odd stamp collecting for mystery diseases—just in case they discovered something important. Fouchier had funding from the program, and he discovered a previously unknown coronavirus in Zaki's sample. Worryingly, it was, like SARS, closely related to what by that time virologists knew were bat coronaviruses.

Zaki posted it on ProMED. A British hospital immediately discovered the same virus in a man with undiagnosed pneumonia who had just been in Saudi Arabia.

Within days, Zaki told me later, the Saudi health ministry had sent an "aggressive" and "threatening" team to investigate his lab. He took emergency leave in Cairo. He was sacked—and informed, he told me, that it was not safe to return to Jeddah.

The Saudi deputy health minister, Ziad Memish, told me it was intolerable that Saudi authorities didn't know about the virus until they saw it on ProMED, three months after the patient died—and just as preparations were at their peak for the biggest annual human gathering on earth, the Hajj in Mecca. That's a real concern: Memish helped run the exactingly thorough Saudi health controls aimed at stopping anything worse than the ubiquitous "Hajj cough" from marring the pilgrimage.

Zaki and Fouchier, though, told me that the upcoming Hajj made it a good thing the virus had been identified so fast, as they had also been able to discover that it did not spread readily. Zaki was convinced that this would not have happened so quickly if he had told only the Saudi authorities.

The virus was named Middle East respiratory syndrome, MERS, as cases were soon found all over the region. It was in local bats, but people got it from camels.

As of November 2019, there had been 2,494 cases worldwide, four-fifths of them in Saudi Arabia, and 858 deaths—this virus has a high death rate. In 2015, a man who had been in the Arabian Peninsula brought MERS to South Korea, starting an outbreak mostly within hospitals: 184 cases, 38 deaths. MERS has appeared in 27 countries, usually as only one or a few cases after someone has returned from travel to the Middle East.

That isn't much spread in eight years. The main reason is that the virus never quite seems to have made itself comfortable in people. MERS can spread from one person to another, but chains of infection die out after a few cases: epidemiologists call it "stuttering" transmission. In viruses that jump to us from animals, contending with our entirely new immune system can be too difficult, and the viruses that manage to get out and into another person might be too few to get much farther.

Moreover, unlike SARS and Covid-19, which latch onto cell-surface proteins in the nose and throat, MERS binds to proteins mostly in the deep lungs. That is one reason it is more lethal than the other two viruses—infections there can kill you. But it also means the virus literally has trouble getting out and into the next victim. Your deep lungs don't cough and sneeze.

The worry with such a virus is that it is under enormous pressure to adapt to its new host, and if MERS does, we could get a virus that spreads more readily but is also deadly. To prevent that, you want to keep its opportunities to infect people, and adapt, to a minimum. This especially applies to protecting patients and staff in a hospital facing a MERS outbreak: even more than SARS, this is a disease of health care, partly because some medical procedures, such as inserting a ventilator into someone severely ill with pneumonia, can expel it from the deep lungs and into someone else. That has been observed with Covid-19 as well.

Hospital infection control staff made huge strides in restricting the spread of MERS, first in Middle Eastern hospitals, then after it invaded South Korean hospitals in 2015. Last year, epidemiologists calculated that increased efforts to diagnose and contain the virus early had averted up to 500 cases since 2016.

So unless it evolves, MERS isn't threatening most of us any time soon. But it is worth knowing about for three reasons. One, it shows that China isn't the only country that doesn't like nasty new diseases being discovered on its territory, or involving foreigners in the response. In years of covering infectious disease, I have encountered all too many examples. BSE, or mad cow disease, started in Britain, and despite science showing it had to be in Continental Europe, countries there denied it for years—even though we knew by 1996 that it caused a devastating disease in humans. When I reported the science in 1997, there was uproar in Belgium, and scientists were forced to lie to support the official denial.

Two, as the only threatening human coronavirus left in circulation after SARS was stamped out, MERS was actually the subject of some coronavirus vaccine work when Covid-19 hit. Those experimental vaccines are now being adapted.

Three, if SARS wasn't enough, the appearance of MERS definitely demonstrated that we should have been working more urgently to prepare for coronavirus outbreaks. How many warnings did we need?

In fact, we got a third one. In 2016, piglets started dying on farms 60 miles from Foshan, the town in Guangdong where SARS is thought to have started. It was called SADS: swine acute diarrhea syndrome. Virologists isolated a coronavirus and found it was 98.5 percent identical to one found in the droppings of horseshoe bats in a nearby cave, the same species linked to SARS and Covid-19. The pigs had probably eaten some. Nearly 25,000 piglets died, and the disease broke out again in 2019.

There weren't any infections among the farmers. But in September 2019, scientists at Zhejiang University in Hangzhou found that SADS can infect cultured cells from humans. So here was another bat coronavirus, it was killing mammals, and it might well infect people. Yet we hadn't done much to guard against these things until we were locked in a global struggle against Covid-19.

Except for a woman named Shi Zhengli, at the Wuhan Institute of Virology, and the EcoHealth Alliance. They've been chasing coronaviruses to where they live—in bats. And that could hold the key to finally getting a grip on these viruses.

## CHAPTER 4

# Don't Blame the Bats

> "We have met the enemy, and he is us."
> —*Pogo*, a comic strip by Walt Kelly

The Covid-19 virus comes from bats. So did SARS. So do MERS, Ebola, Marburg, Nipah, Hendra, and Lassa viruses. So does hepatitis C, which an estimated 71 million people are living with worldwide. After biting the head off what he had assumed was a rubber bat a fan threw onstage during a 1982 concert in Iowa, heavy metal singer Ozzy Osbourne needed the long, painful series of injections required to prevent rabies—another bat virus. (Today the treatment is a little easier.)

And those are just a few of the viruses living in bats that we know cause human disease. In April 2020, researchers reported six kinds of coronavirus previously unknown to science in bats in Myanmar. That adds to the 400-odd found already in China. In 2017, a survey of all known gene sequences of coronaviruses found there were a hundred "clusters," essentially family groups, of the viruses. Ninety-one of those live in bats, making bats the world headquarters of coronavirus evolution. And they carry other kinds of viruses too.

If we want to understand this pandemic, and what we need to do to stop the next one, the connection between bats and viruses needs to be explored, for three reasons. One, if we're going to even keep a watch for the next pandemic, we're going to have to work out what exactly is going on with bats and all these viruses. Two, we must learn which of their viruses might jump to us and take measures to prevent and prepare. And three, and most importantly, we must generally learn how to act on this kind of information. We had it for Covid-19, and we didn't use it.

That's right. A lab in China, in work confirmed by virologists in the US, found a virus very similar to the one that causes Covid-19 in bats in 2013—a full seven years before this pandemic swept the world. Both Chinese and American scientists clearly warned that this kind of virus could well cause a pandemic. And yet, no serious action of any kind was taken. It was no one's job to do that. This is one of the things we need to change.

We knew bats in the Americas carried rabies in the 1950s, but no one knew they harbored this plethora of viruses until 1994. That year, flying foxes, a kind of fruit bat, were found to carry a mystery virus that had killed horses—and two of their human caretakers—in Hendra, a suburb of Brisbane, Australia. After that, the more virologists looked, the more they found.

Wildlife scientists, fearing bats would be persecuted because of this, have accused the virologists of disproportionately targeting bats for viral discovery. But a 2017 review of the research showed that, even after accounting for different amounts of research effort, bats were still significantly more likely to harbor diseases affecting humans than any other group of mammals.

Soon after the SARS epidemic took hold in 2003, Chinese scientists started the long hunt for the source of the virus. As we saw in Chapter 3, it was initially found in masked palm civets at a wet market—but only in Guangdong, never in any other wild or farmed civets. In fact, civets infected with SARS got sick, proving they could not be the home of the virus in the wild, where sick animals don't last long.

In 2004, Shi Zhengli of the Wuhan Institute of Virology and her colleagues started looking for the SARS virus in nature. The team wondered if both the Guangdong civets and people got it straight from whatever animal was the real "reservoir" for the virus, the term for a species that can carry a virus and pass it on but does not itself get sick from it.

Shi's team knew bats could carry viruses without getting sick—and "bats and bat products," they wrote, were increasingly turning up as food or traditional medicine in markets in southern China. So they went to bat caves across China and took blood, urine, feces, and throat swabs from dozens of bats of various species. Shi was nicknamed "bat woman" by her colleagues.

Sure enough, there were viruses 94 percent identical to SARS in the insect-eating horseshoe bat, which lives in Hunan and many other Chinese provinces and across Eurasia. The bat viruses were all similar, but had between them more slight genetic variations than the SARS viruses from humans or civets, although the SARS viruses fit within a family tree of all these viruses. That is what you'd expect if bats are where these viruses live naturally and just one or a few of such viruses moved into civets and us.

None of the viruses they initially found was exactly the same as the SARS virus. For one thing, none had the same binding area on the big spike proteins on the outside of the virus, which in SARS latches onto the ACE2 protein on human (and civet and bat) cells—the same receptor Covid-19 uses. But they kept looking.

In 2009, they partnered with the PREDICT program of the US Agency for International Development (USAID). PREDICT was launched after the H5N1 bird flu scare in 2004, which we'll look at later. It sets up local labs and surveillance in countries with "hotspots" of zoonoses, and the EcoHealth Alliance is a major participant. Kevin Olival, now EcoHealth's vice president for research, mostly works in Indonesia and Thailand, countries that can use more help building virology infrastructure than China, which has become one of the world's leading producers of world-class research.

But there is a PREDICT team working alongside Chinese scientists at a forest site in China's southern Yunnan province. It is only 40 miles from Kunming, a city of six million, but it is a zoonosis hotspot, with a well-populated bat cave. Olival told me about their research project.

The procedure, he said, is to trap bats when they flutter out of the cave to hunt just after dark. The trap looks like a giant harp, with two sets of vertical cords strung in an open frame. The echo-locating bats detect the first cords and do a mid-air turn to fly between them—only to become lodged against the second set. No longer able to stay aloft, they slide down into a big, soft bag at the bottom. "They just snuggle up in there," says Olival.

The scientists and technicians already have lights, bottles, labels, and sampling swabs laid out on a folding table nearby. They take a throat swab, an anal swab, and a blood sample from each bat—then the bats fly off to resume hunting. As a conservation as well as disease-hunting organization, says Olival, "we don't want to harm the bats."

The samples from Yunnan are analyzed for coronaviruses, but PREDICT has operations like this looking at different kinds of high-risk wildlife, and different pathogens, in Bangladesh, Brazil, Colombia, Indonesia, Malaysia, and Mexico as well. The findings are analyzed and mapped to predict potential disease outbreaks, and importantly, says Olival, the information is fed back to the communities at risk, so they can protect themselves.

In China, though, the EcoHealth Alliance was joining a disease hunt that was already underway—and the collaboration quickly paid off. In 2013, Shi's lab found two viruses in the Yunnan bats that were 95 percent identical to SARS and had an exterior spike protein with a sequence they knew would bind to the same ACE2 protein on human cells that SARS used to invade us. The two sequenced viruses were dead—many samples from the bats are taken back to the lab in a preservative, as that makes them less risky and easier to deal with.

But they also brought some samples back alive. From one, the team managed to isolate a live virus that could infect both bat and human cells. It was also instantly recognized by antibodies, immune proteins that are highly specific for particular pathogens, from SARS patients in 2003. "Bat coronaviruses remain a substantial global threat to public health," Shi's team concluded.

In 2017, the team reported more SARS-like viruses from the bats and discovered that, like some other viruses, they swap gene segments around. They found segments with all the exact gene sequences required to build the original SARS virus in bats from the one cave near Kunming and evidence that the viruses were actively recombining gene segments. After 14 years, the long search was over: they knew for sure where SARS came from.

But besides SARS, they found a wide variety of slightly different coronaviruses similar to SARS that are able to latch onto the human ACE2 protein to invade cells. "The risk of spillover into people and emergence of a disease similar to SARS is possible," they warned. That, of course, is what has now happened: Covid-19 latches on to ACE2.

Meanwhile, virologist Ralph Baric and his team at the University of North Carolina reconstructed one of the viruses Wuhan had discovered using the gene sequences and found it infected human airway cells in culture just as well as the SARS virus—the Urbani strain. It made mice with human ACE2 proteins sick. Yet it was different enough that an experimental SARS vaccine didn't protect them—showing that even if we beat one type of coronavirus, even very similar ones could be entirely new challenges. The title of a 2015 publication of the work warned that SARS-like bat coronaviruses "show potential for human emergence," and the report spoke of "a need for both surveillance and improved therapeutics against circulating SARS-like viruses."

In 2016, further work called one of the viruses "poised for human emergence." "The virus has significant pathogenic potential," Baric and the team concluded—and if it emerged, we had no vaccines.

So we knew there were viruses like SARS out there that could infect humans and cause illness without having to spend time adapting in another species first. We knew this seven years ago. And since then, more research has only confirmed that. It was even reported in the press: at that meeting in 2016 in Vienna on emerging diseases where people were worried about Nipah, Kevin Olival told me, and I reported, that PREDICT had helped find "a Chinese virus closely related to SARS but different enough that prototype SARS vaccines won't work against it."

It got even more alarming. In 2018, Shi's team reported that the viruses were already trying us out. They found antibodies to the bat coronaviruses in people living near the caves in Yunnan, showing they had been infected—and they had not been exposed to SARS in 2003, or traveled. There were also antibodies to SARS-like viruses in market traders in Guangdong in 2001, well before SARS exploded. But those weren't found until 2004, after SARS was gone, in a retrospective analysis of stored blood samples. With SARS we were caught blind, but with Covid-19, we knew these viruses were checking us out before this pandemic even happened.

"It is highly likely that future SARS- or MERS-like coronavirus outbreaks will originate from bats, and there is an increased probability that this will occur in China," Shi wrote in a review of the research just last year. "Therefore, the investigation of bat corona-viruses becomes an urgent issue for the detection of early warning signs." Every disaster movie starts with someone ignoring a scientist. It is now too late for warnings.

Perhaps the team's saddest paper came out on January 29th, the same day my *New Scientist* piece was headlined that the new coronavirus looked set to go pandemic. This time, all the authors were Chinese scientists, most in Wuhan, and the new disease was raging in their town. They recapped the work to date: they had discovered SARS-like coronaviruses in their natural reservoir, bats. "Previous studies have indicated that some of those bat SARS-CoVs have the potential to infect humans," they recalled. The news this time was that the new virus killing people in Wuhan was 96 percent identical to one of those bat viruses, RaTG13, and used the same cellular receptor, ACE2.

We warned you, in other words. But good scientists that they were, they got on with what to do now. "Future research should be focused on active surveillance of these viruses," they wrote, and broad-spectrum drugs and vaccines should be developed against this group of viruses in general. "Most importantly, strict regulations against the domestication and consumption of wildlife should be implemented."

That last comment gets us to the most important question raised by all this. Bats live everywhere. Why have these bat viruses broken out in humans, twice, in China? Is it the bats? Or the way people interact with bats?

It is actually hard to catch viruses directly from bats. Only six

of the 218 people who lived near the bat caves in Yunnan had antibodies from infection with bat coronaviruses, even though they
regularly saw the bats near their homes. Similarly, MERS has been
found in Saudi bats, but humans only ever catch it from camels,
which apparently carry the bat virus with no ill effects. As mentioned earlier, poor little Emile Ouamono in Meliandou in Guinea
caught Ebola from a bat and died, triggering the Ebola epidemic
in West Africa in 2014. But children in his village routinely caught,
roasted, and ate those bats with no apparent problem, says wildlife
virologist Fabian Leendertz, who led the expedition to Meliandou
to try and figure out what happened. He doesn't know why Emile
was unlucky.

John Mackenzie of Curtin University in Australia (no relation to
the author) told me that no one there has ever caught Hendra virus
directly from a bat—only from horses, which might get it by eating
the fibrous remnants of fruit the bats spit out or the afterbirths
shed from birthing roosts. Catching Nipah requires either a pig
intermediary or sharing a drink of palm sap with a fruit bat. Australian wildlife activists regularly nurse injured bats back to health,
and only two, says Mackenzie, ever got Australian bat lyssavirus, a
virus closely related to rabies carried by Old World bats. Sadly, they
died; nowadays, bat rescuers are all vaccinated.

Confused rock stars aside, people have gotten rabies by handling bats in the Americas. But today, Britain and Australia are
considered rabies free, even though their bats carry lyssavirus. Bats
just don't pose enough risk to people to be counted the same as
other rabid species, like dogs or raccoons.

I know a woman in a picturesque Cotswolds town in England

who rescues bats. She has a roomful of cages and baskets sheltering convalescent bats of nearly every species in Britain, some endangered. She freely handles them, feeds them, bandages wounds. Only one British species, Daubenton's bat, is known to carry bat rabies, she assured me on a visit, expertly lifting one out of its basket. They're cute little things, with incredibly soft brown fur. I trust her judgement, but I was happy to leave handling them to the experts.

So how have SARS and Covid-19 gotten into us? Blame has fallen on the wildlife trade—especially as both diseases emerged in winter, the season for hunting and slaughtering animals in agricultural societies, so by extension the season when eating game meat has been traditionally considered good for one's health in China.

In April 2020, the executive secretary of the UN Convention on Biological Diversity called on China to shut down wildlife markets, like the one in Wuhan linked to Covid-19. "The message we are getting," said Elizabeth Maruma Mrema of Tanzania, "is if we don't take care of nature, it will take care of us."

Opinion is now divided, though, on what role that market played in Covid-19. Two-thirds of the cases China first reported on January 24th had links to it, and the rest didn't. If the virus came from some animal sold on the market, I still hear people asking, why were a third of cases not linked to it? Possibly because many early cases got the virus from other people, not an environmental source, and it just happened to spread to the market early—markets are, after all, sources of human contact.

"I am strongly of the belief that the virus as we saw it in Wuhan is pretty much exactly the virus that was in bats—it just happened to have everything it needed to spread in humans," says Rambaut. "I think the market cases were just part of a bigger cluster. It doesn't mean the market was the source." But the association with the market stood out, maybe because people associated markets and SARS. And in January, only people with an association to the market, or another case, could be tested for the virus—we don't know how many with no association there were.

And just as civets proved to be at best incidentally related to SARS, so it seems are pangolins and Covid-19. Related viruses were found early on in the scaly creatures, the world's most-trafficked, highly-endangered mammal, and Chinese scientists proposed that pangolins were the intermediate host that spread the virus to people.

"The pangolin turned out to be a red herring," says Rambaut. The SARS-like viruses found in them are far less like the Covid-19 virus than are viruses found in bats.

RaTG13, the bat virus that was closest genetically to SARS-2, the Covid-19 virus, wasn't identical to it. "We estimate they split from a common ancestor between 40 and 70 years ago," says Rambaut. But other features of the viruses suggest that "the lineage that gave rise to SARS-2 was in bats for almost all of that time. I don't think we require an intermediate host to explain any of the features of the SARS-2 genome."

There are two unusual bits of gene sequence found in the Covid-19 virus, and as of May, none of the known bat viruses had them. One of those did turn up in a pangolin virus—but "it's possible both of them are in a bat virus in some combination," says Rambaut. The variety of viruses in bats is huge, and it took 14 years of dogged bat sampling to find viral gene sequences that precisely matched the 2003 SARS virus. It is not surprising that researchers haven't found Covid-19's dead ringer yet.

Yet accounts of SARS nearly all state with assurance that civets were the intermediate host, despite the lack of evidence the virus even needed one. The same kind of uncritical tale of Covid-19 and pangolins is now becoming entrenched. If that leads to further persecution of pangolins—already at severe risk due to their use in traditional Chinese medicine—it would be tragic.

So if the virus came straight from bats, but it's hard to catch viruses from bats normally, how did we get it?

If it did make that first jump at a market, there is hope of stopping this from happening again. China shut down live wildlife markets across the country in late February, and activists hope it will follow that up with a permanent ban. That was supposed to happen after SARS. According to TRAFFIC, however, Guangdong imposed a trade ban on wildlife meat in late April 2003, but lifted it by mid-August—after SARS had disappeared—for 54 captive-bred species. Business as usual rapidly resumed.

That may happen again. By late March, with cases of Covid-19 in China falling after weeks of lockdown and the problem perceived as over, China's wildlife markets were reported to be re-opening.

In any case, what are bats doing there? Peter Li of the University of Houston–Downtown says eating exotic wildlife is not traditional among the vast majority of Chinese. He says rural families turned to catching and then raising wild animals as a way to get food and then to make money, after the upheavals of the 1960s in

China. Since then, a wealthy and powerful industry has emerged selling ever-more exotic meat to China's many wealthy city dwellers. "Demand for wild animal meat from the consumers is false," says Li. "The demand has been created by the traders and restaurant owners who claim it is good for health, longevity, sex, brain health." That suggests it may be possible to reverse the fashion, and reportedly it is less popular with younger Chinese.

It isn't just fashion, though. Bats are traditionally eaten in southern China, as well as elsewhere in southeast Asia and, indeed, in Africa—smoked bats are popular in Ghana, and Leendertz says they have become common game all over Africa as larger animals become rarer. However, people tend to eat the meaty, large-bodied fruit bats, not tiny insect-eaters like the horseshoe bats that carry SARS-like coronaviruses.

But maybe we're on the wrong track by thinking of bats as food. Horseshoe bats are used for TCM, traditional Chinese medicine, which is very widely used in China. The WHO reported in February 2020 that herbal medicines prescribed by TCM figured prominently in China's medical response to the Covid-19 epidemic.

Ye Ming Sha—night brightness sand—is dried, powdered bat feces. It's not hard to find: type it into a search engine, and you'll turn up numerous online sources. One, charging $12.38 per 100 grams, lists one of the source species as horseshoe bats. "It cools the Blood, reduces Stasis and stops pain. It treats eye disorders…malarial-like disorders, childhood fright, painful urinary dysfunction, vaginal discharge, scrofula and swollen sores," the product description reads. A happy customer in February posted, "already put into use."

But mostly, Ye Ming Sha is used for eye problems. The *Clini-*

*cal Handbook of Chinese Prepared Medicines* says it "clears heats, nourishes the eyes, and improves night vision (due to high levels of vitamin A)." An online site about TCM explains that "bats are blind, fly at night" so their droppings are good "for vision, especially at night." (In fact, bats have excellent vision for the same reason birds do: they fly.)

Sampling in Yunnan found bat coronaviruses in fresh horseshoe bat feces. Drying fresh bat droppings collected in the wild might kill viruses in it, but possibly not always completely, especially if the resulting powder is applied to an unprotected body part: there are ACE2 receptors in the eye, and evidence suggests the Covid-19 virus is especially persistent there, and that eyes might be a major route of infection. When I asked TCM practitioners online, they advised applying a water extract of Ye Ming Sha directly to the eye.

According to information collected by TRAFFIC, dried bodies of horseshoe bats are also a folk remedy for cough—ironic, given Covid-19's signature symptom. Perhaps the greatest risk is not from the remedies themselves, but from, and to, the generally impoverished people who catch the bats or collect fresh droppings. The bat guano used for fertilizer is often from old deposits and may pose less risk. If collectors are infected with a bat virus, they could then spread it to other people, maybe while delivering bat products to a market—one possibility for the so-far elusive patient zero in Covid-19. Horeseshoe bats live in Hubei and indeed are widespread in China.

It is not unreasonable to think that China might reduce the risk of zoonotic disasters by cleaning up wet markets, whether or not those were the source of Covid-19. We know for certain that they are the source of some worrying strains of bird flu. Peter Daszak says it would make sense to at least improve biosecurity there. Currently, the large number of species sold in wet markets are stacked in cages, freely passing body fluids and any accompanying viruses among themselves, bats included. Better hygiene might reduce that.

At least one might hope there are better ways of cleaning up markets than were used in April 2020 in Surakarta, Indonesia. After news of the origin of Covid-19 got out, live fruit bats—who do not, as far as we know, carry these viruses—were taken from a wet market and burned alive.

But controlling wet markets is one thing. TCM is quite another. It is held in high esteem in China. Much of it is undoubtedly valuable: artemisinin, currently the world's leading antimalarial drug, was derived from a traditional herbal treatment by one of the great women of science, Tu Youyou, who won a Nobel Prize for it in 2015.

All the same, some Chinese think it might be time to reconsider some TCM ingredients. The *Chinese Pharmacopoeia* is the authority for materials approved as medicine in China. The current 2015 edition includes bat feces, but a new edition is due out in 2020. "There has been a lot of pressure exerted on the Chinese authorities to remove the bat dung as an ingredient from TCM in the *Chinese Pharmacopoeia*," says Peter Li. "At this point, I don't have confirmation that it will be removed."

The larger issue, though, is conservation. If maintaining biodiversity in the wild reduces our risk of zoonosis, as scientists say, then all of China's wildlife trade, not just markets, needs careful consideration. The use of animals for medicine often bypasses wildlife markets, but is driving some species to extinction. And some uses could stem more from commercial claims than real traditional cures.

Knowing all this about bat viruses and their transmission, one key question remains: Why bats? It seems humans get these viruses from bats, but how and why do bats get them?

For years this has been explained with a much cut-and-pasted set of proposed explanations: bats are everywhere, some live in large communities, they travel long distances. But those things are true of other species, including our own, and we don't naturally harbor things like Ebola or Covid-19. Living in large colonies facilitates the spread of bat diseases to other bats: witness the white nose fungus severely threatening some bat species in North America. Why should that also facilitate things that kill people?

It now seems more likely to be a result of none of those factors, but instead, something unique to bat biology. Understanding this helps to explain what we have to do to stop pandemics like Covid-19 from happening again. Spoiler alert: it absolutely is not killing the bats.

Nearly a quarter of all mammal species are bats; only rodents account for more. Bats are the only mammal that truly does powered flight, using its muscles for lift rather than simply gliding like flying squirrels. In terms of evolution, this was a spectacular success.

Flying meant bats could occupy many niches—spaces in the environment that supply the shelter, food, and mates they need— that nothing else could occupy. Lots of niches meant lots of species, and two wildly different tribes evolved: the big vegetarian fruit bats of Eurasia, Africa, and the Pacific; and the little, echo-locating insect eaters of pretty much everywhere but Antarctica.

But flying has a downside: it takes a huge amount of energy. A bat's heart can beat 1,000 times per minute. While they fly, they burn sugars and other fuel for energy and consume oxygen just as we do when we exercise—but while flying, a little insect-eating bat does that at twice the rate of a similarly sized mouse running flat out.

All these chemical reactions generate damaged molecules, called free radicals, that are highly reactive, like fires in your cells. Bats have extremely efficient systems for snuffing these out. This has a useful side effect: long life. Free radicals are thought to cause many of the changes of aging and may be why smaller animals, with higher metabolic rates and more free radicals, live shorter lives. But whereas a mouse lives two years, a bat of the same size with an even higher metabolic rate—but a very good free radical extinguisher—can live for 40.

There's another side effect, though. Bats' high energy turnover produces other molecular fragments, bits of DNA. Such fragments aren't damaging in themselves, but in us, they can only mean one thing: infection, by some pathogen that left its DNA lying around. So in humans, any such fragments trigger massive inflammation, an immune reaction that kills cells infected with viruses. However, in a bat, DNA fragments are normal, and a bat would just damage itself if it unleashed inflammation on cells that have them. So bats dial down inflammation. That means they need another way to protect themselves from infection.

To do that, they evolved a different method of fighting viruses: they don't. Instead, they mount a kind of nonviolent exclusion.

In her free time, Cara Brook of the University of California at Berkeley works as a science educator and conservationist in Madagascar. But her day job is virology, and in February 2020, she published work in which she infected bat cells growing in dishes with Ebola and Marburg viruses, two human pathogens found in fruit bats. The cells mounted a lightning response, rapidly turning on a slew of genes that effectively stopped the viruses from invading the cells.

Whereas humans would turn on a complex cascade of inflammation reactions to clear out cells infested with a virus, bats try to stop viruses from ever entering their cells at all. In Brook's dishes, a few cells didn't react fast enough, so they were infected, the virus could replicate, and the infection smoldered on—but it was held at bay as most of the cells stayed virus-free. Such a low-level infection, Brook calculates, can last a bat's whole life, even if it is a relatively long one.

That's why even though many families of viruses live in bats, only one, rabies, seems to cause much disease. Many of the symptoms of our illnesses are triggered not by what the virus is doing to us, but by the efforts of our immune system to kill the virus. That's why so many diseases, including Covid-19—and of course flu—start with the same, infamous "flu-like symptoms." Bats don't do much inflammation at all, and they stop most viruses from doing too much direct damage, so they don't get sick like we do from them.

But the viruses, like any evolving organism, fight back. The ones that attack just a little bit faster than the others are the ones that get into an occasional bat cell and replicate. Those viruses become more numerous—or, in evolutionary terms, aggressive infection is selected for. Brook suspects this is what makes bat viruses extra deadly in us: they have evolved to beat the bat's hair-trigger response, so in us they move ferociously fast.

In addition, bats' high metabolic rate means they normally have a higher temperature than humans. One of our defenses against viruses is to run a fever, as our viruses are often hurt more by a slight increase in temperature than we are. A fever might just make bat viruses feel more at home in our bodies.

The mechanisms bats use to coexist with viruses could teach us a lot about how we might control our own viral infections, says Kevin Olival. And the fact that damping inflammation seems to contribute to bats' longer life spans—and might also be what stops them from getting cancer—could teach us even more. Meanwhile, apart from studying them, the best thing to do with bats is to just leave them alone.

That may seem counterintuitive. If we want to protect ourselves from viruses that normally live in bats, shouldn't we just get rid of bats? Unfortunately, people do routinely destroy bat colonies out of fear of disease, especially rabies, even though disrupting bats is more likely to spread the diseases than stop them, as the uprooted bats that escape fly everywhere. There are already reports of people destroying bat colonies in misguided efforts to fight Covid-19.

More to the point, though, "You can't get rid of bats," says Olival. "The world needs them." They are often "keystone" species on which many others in an ecosystem depend.

Hundreds of species of fruit, for example, depend on bats for pollination, including mangoes, bananas, and guavas—and if fruit doesn't rock your world, you should know that bats are also essential to the agave cactus used to make tequila. The vital baobab trees of African savannas are exclusively pollinated by bats. Insect-eating bats—like the ones that host the Covid-19 virus—can eat their weight in insects every night, especially nursing mothers. Those meals include disease-carrying mosquitoes. The bats also eat tons of moths that are major crop pests. Bats are calculated to do $3.7 million worth of crop protection a year in the US alone, no polluting pesticides required. Losing them would trigger domino effects through the farming ecosystem that would cost even more.

Fruit bats—like the ones thought to host the Ebola virus—are vital for dispersing seeds in tropical rainforests. "I often say, no fruit bats, no rain forests," says Andrew Cunningham, a veteran wildlife and animal disease expert with the London Zoological Society. "In fact, given the role of rain forests in carbon storage and weather patterns, you could take this further to its logical conclusion and say no fruit bats, no humanity as we know it."

"Simply left alone, bats are harmless and highly beneficial," says Bat Conservation International. Of course it would say that—but in 2006, a team of scientists at the Arthropod-borne and Infectious Diseases Laboratory in Fort Collins, Colorado, agreed.

In a review of the research, they concluded that bats are critical for nearly all biological communities on land. "Myths and misunderstandings...have led to efforts to extirpate bat populations, with serious consequent effects on insect control and crop production, without coincidental reduction in the already low incidence of rabies virus transmission by bats."

This holds true for other viruses as well. "There has been public and political pressure in Queensland to manage Hendra virus by culling or dispersing fruit bat populations," Australian scientists reported in 2015. But, they found, the amount of virus in a population of bats didn't depend on their population density, so reducing that density would not reduce the virus. Stress on the bats, though, would increase it. In 2008, researchers found that hunger made Hendra virus more prevalent in flying foxes than any other stress, making the steady loss of trees where fruit bats can live the biggest risk. And climate change and wildfires just cause more of it.

The 2015 report suggested that restoring forests of wild, fruiting trees to entice the bats away from people and horses would be the best way to prevent Hendra. "Bats are not the problem. They don't cause disease emergence," says Cunningham. "People do, by destroying and encroaching on their habitat and by catching, trafficking and butchering them. This can even infect other animals nearby that, if infected, might be able to carry and even multiply the bat virus, further increasing the risk."

In any case, says Olival, besides being ecologically appalling, eradicating bats would simply be impossible—there are too many, and they fly. And then the remaining bats might just carry more virus. When a cave full of bats carrying the deadly Marburg virus was smoked out in Uganda, it was rapidly re-colonized by young males from other colonies, carrying more Marburg than the originals: Marburg is a childhood infection in bats.

The problem with bat viruses, the researchers note, is not the bats: it is that when one of their viruses does jump to us, we let it get away. In West Africa in 2014, there was one transmission of the Ebola virus from a bat to a human child—and thousands of transmissions between people after that. Covid-19 started with one jump of a bat virus to one or a few humans. Then that was followed by millions of transmissions among us. That second thing—the transmission of virus between humans—is the problem.

The answer, says EcoHealth Alliance, is both surveillance, to spot and contain diseases early when they do reach humans, and conservation, to maintain intact ecosystems where bats are unlikely to encounter people or move into farms or towns. If nothing else, surveillance is cost-effective. Over 10 years, says Olival, PREDICT costs around $200 million, much of it to establish ongoing capabilities to monitor emerging infections in 30 low-income countries. That is a tiny fraction of the trillions in emergency relief the US has allocated as a result of Covid-19, and that in turn is only part of what the pandemic has cost.

But PREDICT also illustrates the problems with the surveillance we have so far. The program's funding came to an end in 2019, and fieldwork stopped when the money ran out in September. It was granted another $2.26 million to continue another six months from April 1st, because the labs it helped set up are, in some countries, the only labs that can detect the Covid-19 virus and, without PREDICT funding, trained staff will likely be lost. PREDICT labs were the first to detect people with Covid-19 traveling from China to Cambodia and Thailand.

But that funding was only guaranteed through September 2020. This kind of capricious funding, dependent on varying levels of interest or capacity in the scientific or political world, has prevented the serious, day-in, day-out surveillance required to prevent pandemics.

PREDICT has at least built countries' local capacity to continue monitoring their own viruses, says Olival: "We don't fly in, collect samples and fly out." The scientific capability they leave behind might be among the best legacies of the program. David Heymann, who headed the WHO's campaign against SARS, believes that is what the world most needs to catch the next pandemic virus that emerges.

The question is: What did we do with the warnings we got from PREDICT? The viruses they helped collect in Yunnan allowed Shi and Baric to warn us about SARS-like viruses that could emerge in humans with no further alterations needed. The warning was taken seriously enough that the US decided to renew its funding of PREDICT's work in 2019, including work to investigate the viruses. That hit a major snag though, in April 2020, which we'll discuss in the next chapter.

What about an actual response aimed at protecting ourselves from these viruses, though? The aim of the WHO's R&D roadmap is to develop vaccines, treatments, and diagnostics for its list of priority pathogens, which includes coronaviruses in general. In theory, we could have done that. In practice, until one of these viruses causes extensive disease in people, much expenditure is unlikely, whatever the WHO roadmap says. But we could at least have developed a PCR test that distinguishes between SARS-1 (or as some virologists are calling it, SARS classic) and the related bat viruses in case one did emerge, and put more effort into surveillance for any coronaviruses in humans. The bottom line is: we didn't even do that.

Perhaps if Dr. Ai in Wuhan had had a more specific test on hand and realized that her patients didn't have SARS, but something new, the alarm would have been greater and the response faster. EcoHealth and other organizations promote the concept of One Health, communication and coordinated research and monitoring of disease between researchers and clinicians who deal with human and animal health. It's a sensible idea.

But it will accomplish little as long as no one in government is tasked with the job of using this information to fund the precautionary responses that would make us safer. Or an intergovernmental forum, come to that. We will return to that later.

It's clear: we had warnings and we didn't act on them. However, there's one disease where we *have* taken the warnings into account, for which One Health thinking and pandemic planning are well advanced: good old reliable flu.

## CHAPTER 5

# Wasn't the Pandemic Supposed to Be Flu?

"In the year of 19 and 18, God sent a
mighty disease.
It killed many a-thousand, on land and on
the seas."

—Blind Willie Johnson, "Jesus Is Coming Soon"

In January 2004, I went to a meeting at the venerable Royal Society in London, called to take stock of what we had learned from
the SARS nightmare that had ended six months before. The coffee
break arrived, and people from the conservation groups were chat-
ting in hushed tones about civets. It got depressing. I headed for
the coffee in the back of the room.

There I saw someone I did want to talk to. Ab Osterhaus is one
of Europe's top virologists. His lab had just "done Koch's postulates"
on the SARS virus, a rarely met standard for proving a pathogen

causes a disease. And he was leaning against a pillar, looking very, very shaken.

I wasn't sure I should say anything, but Ab's a pretty informal Dutchman, so I asked if he was okay. He told me he had just been exchanging emails with colleagues in Hong Kong. "It's this H5N1 bird flu," he said. "If this adapts to humans, it could be really bad." He searched for a word. "Civilization ending," he said.

In early 2004, Ab wasn't the only flu expert who was very worried about H5N1. In fact, those experts still are. Yes, the Covid-19 pandemic is a coronavirus, not a flu. They are quite different. But we're talking here about pandemics in general. It is to be hoped that, having seen the Covid-19 pandemic, we handle the next flu pandemic better. It would be only fair, because the last flu pandemic messed up the way we are handling Covid-19.

Flu, influenza A to be formal, is the one pandemic we know is coming. We know other diseases can go pandemic—and if anyone had any doubts, Covid-19 ended them. You can debate, perhaps, the pandemic potential of some of the viruses on the WHO's priority list. But flu is a different story. Pandemic is what flu does. You can't talk about how pandemics happen and how we respond to them without understanding flu.

First, flu 101. Stick with me for a moment, you'll see why this really matters soon. The virus consists of eight chunks of RNA, coding for a mere 11 proteins, and a shell studded with two of those proteins, hemagglutinin and neuraminidase—thankfully abbreviated H and N. These come in different varieties, which have numbers, and paired together, they identify the type of flu virus. Right now, two kinds of flu, H1N1 and H3N2, are circulating in humans. But flu viruses that circulate in ducks, the original host of flu, carry 16 different kinds of H and nine kinds of N. Plus there are two more of each unique to—you guessed it—bats. Like most of the other varieties, they leave us alone.

Flu viruses adapt to particular hosts—our current two kinds of influenza A are adapted to us and don't infect birds. (There's also an influenza B, which circulates with the two influenza A viruses and makes people sick every winter, but it never seems to go pandemic, so let's mostly ignore it here.) Likewise, bird flu viruses are adapted to birds, and don't—normally—infect us. Both bird and human flu viruses can infect pigs as well as the pigs' own kinds of flu, and humans can catch what emerges.

The flu virus is transmitted in the droplets we exhale, like Covid-19. Droplets evaporate and fall to the ground quickly in warm weather, so flu does better in cool weather. Derek Smith of Cambridge University and colleagues worked out how this leads to the annual flu epidemic. Accidents of geography mean that in East and Southeast Asia, cool rainy seasons are always happening somewhere, at different times. So there is always a flu season happening somewhere in the region, and flu is constantly infecting people and evolving.

Then, as winter sets in in the Northern Hemisphere, the flu in East Asia breaks out and circles the globe. Then it does the same in the Southern Hemisphere's winter. Flu basically mounts its own pandemic every year, except we don't call it that because it is routine.

## Wasn't the Pandemic Supposed to Be Flu?

The viruses that dominate this yearly tour of the planet are the ones that can dodge our immune systems best and get into the next human quickest. To get to the top of this class, flu plays a crafty game. The big H protein on its surface attracts most of your immune system's attention, and it constantly mutates, at seven different hotspots. It eventually accumulates so many little changes that many of your immune defense proteins, antibodies, that would recognize and attack the last flu you got don't quite recognize this virus. So you get sick again.

This is easy for flu: it makes a lot of mistakes when it copies its genes because it doesn't have an enzyme for fixing them. The virus that causes Covid-19 does have that enzyme, so it has at least started out this pandemic with what seem to be much more stable genes—although it's still an RNA virus, so it can evolve fairly quickly if it's pressured to. Mutation is random—and in RNA viruses, even relatively stable ones, fairly frequent. When a random mutation happens to make its bearer survive and reproduce better than others that don't have the mutation, bearers of the mutation can become more successful and numerous. That is not random. That's evolution.

With flu, your immune system still recognizes the unchanged bits of the slightly evolved H and N and the rest of the virus, so you can mount some immune response and keep the infection in check. That's why most ordinary flu is just that, ordinary: just enough to make most of us miserable for a few days each winter, and—usually—no worse.

This constant change is also why we need a new flu shot every year. Ordinary winter flu isn't always trivial. Just like Covid-19, it is far more lethal in the elderly and people with underlying conditions like diabetes. That's why health agencies recommend a flu vaccine for such people every year. The vaccine you get next autumn needs to immunize you to the flu that will circulate the following winter—which will be a little different from the flu that circulated this past winter. But it takes six months to grow enough flu virus to make that vaccine.

So twice each year, vaccine companies and flu virologists come to a meeting at WHO headquarters in Geneva and try to predict what precise flu virus will be circulating in a bit more than six months' time, so they can start growing vaccine. They hold one meeting for the Northern Hemisphere, one for the Southern.

It's not easy. The guesswork is based on years of sophisticated observation and scientific analysis. Even so, flu sometimes pulls a surprise, and the vaccine virus the companies spent six months growing turns out to have been different, immunologically, from the flu virus that ends up dominating that flu season. Or they guess right, but growing in eggs sometimes causes the vaccine virus to evolve and adapt to eggs, and what comes out isn't quite what they put in. In Australia in 2017, the H3N2 virus in the vaccine did this and provided little protection. It's not the world's greatest vaccine technology.

There are other flu vaccines. There are live, weakened flu viruses you take as nose drops, and normal flu vaccines grown in cell cultures instead of eggs. But there are only a few such vaccine factories. There just isn't enough money in flu vaccines to justify much investment. Not everyone bothers getting vaccinated for a disease that in many is mild. Even if they do, it's only once a year, and the companies can't charge a very high price or they'll lose what customers they have. A few years ago, a vaccine maker pulled the plug on a plan to build a new flu vaccine plant in the US: they just couldn't make it add up economically, even with substantial financial support from the US government.

Flu experts have been warning for years that we need to fix this situation, because every now and then flu makes a really big genetic change and outfits itself with an H and an N that no one, or almost no one, has ever encountered before. Then, whatever immunity to flu we got from viruses in recent years doesn't work, especially if the new virus is substantially different. Such viruses cause more severe disease, and we put up so little fight against them, they can circulate in seasons other than winter. We call this global epidemic of flu a pandemic.

That happened in 1918 when a particularly deadly flu virus started circulating. You will have heard about it, perhaps because of its recent 100th anniversary or perhaps because so many people are making comparisons to Covid-19. It was called the Spanish flu, because it started during World War I and news of it was censored in the countries that were fighting—but not in Spain, which wasn't fighting. It was lethal: there were tales of people getting on buses or trains feeling not too ill and dying before they arrived at their destination. Opinion is divided, but at least 50 million dead is a safe guess—in a world that had a quarter of the population of today. In any case, it caused more deaths than the war itself.

The virus was more aggressive than most flu in attacking the deep lungs and causing pneumonia directly, and also triggered bacterial pneumonia, all of which Covid-19 does too—except in 1918, there were no antibiotics for the bacteria. The pandemic may have helped end World War I—and start World War II. Its third wave in April 1919 took the most conciliatory negotiator, US president Woodrow Wilson, out of treaty talks in Versailles, helping lead to a treaty so punishing for Germany that it is often blamed for the rise of Hitler. The US lost 675,000 people to flu in the 1918 pandemic, more than in World War I, World War II, the Korean War, and the Vietnam War combined.

Incredibly, we have actually been able to analyze its structure. Virologists recovered the virus from an Inuit woman who died of it and was buried in permafrost and reconstructed it in 2005. There is still some disagreement about where that virus came from and where it first broke out. Some experts think it was a bird flu that managed to adapt to humans.

Some think it got some genes from earlier human flu. If two flu viruses invade the same cell, their eight RNA bits replicate and then reassemble in random mixes. If a bird flu and a human flu invade the same cell, some of the viruses that emerge can have totally new H and/or N proteins, as well as other components, a mash-up to which we have much less immunity than we normally do to flu.

History records flu pandemics back to 1510. The one in 1918 seems to be the most lethal on record. But by 1921, that same virus was ordinary winter flu, not because it mutated massively, but because most people had encountered it, survived, and developed some immunity. It proceeded to circulate every winter until 1957, when it swapped its H and N for replacements from a bird virus— which virologists named H2 and N2, because this was all news to them and these were only the second Hs and Ns they'd seen.

That pandemic was called the Asian flu, and it killed about two to four million people—a lot compared to the 250,000 to 500,000 thought to die worldwide in a normal flu year. In 1968, that virus swapped its H2 for what we (of course) called H3, also from a bird—"only" a million died in the pandemic of what was called the Hong Kong flu, as the change in the virus was not dramatic enough to completely defeat our existing immunity. Both viruses are thought to have evolved in southern China, which agrees with Derek Smith's findings about flu coming from East Asia.

Meanwhile, back in 1918, the pandemic virus also killed many pigs, but then they developed immunity like we did. It kept circulating, as agriculture modernized and hog herds mushroomed in size. Then in 1998, swine flu picked up genes from flu viruses normally found in people and in birds, both of which can also infect pigs—and hybridize with the pig viruses.

That aggressive new "triple reassortant" virus dominated North American pig farms within a year. In 2004, virologists warned these viruses had pandemic potential because they were also infecting occasional farm workers and sometimes picked up H and N proteins humans weren't used to. By this time, there were two flu viruses circulating in people: the H3N2 virus that had taken over in the pandemic in 1968 and a fairly mild descendant of the 1918 H1N1 virus, which we think escaped from a lab somewhere. (Seriously. No one quite knows where. Maybe Russia.)

On the 21st of April, 2009, the US CDC reported two children in California with H1N1 flu—except this wasn't the mild, human strain of H1N1. This was more like the H1N1 virus found in pigs, except the children hadn't been near any pigs. Then Canada issued a travel warning about a flu outbreak in Mexico—unusual in April—that had already killed at least 60 people. The US found two more kids with swine flu in Texas. It all made ProMED. The morning of the 24th, I emailed my editor at *New Scientist*: "This is exactly what an emerging pandemic would look like."

It was. Five days later, the WHO declared a pandemic was imminent. We knew this because a new virus was spreading human-to-human in North America that no one seemed likely to have much immunity to—the US CDC was conducting daily briefings for us health journalists and (after a few pointed questions) said as much. That meant it was only a matter of time before it was spreading rapidly on another continent. When that happened, the WHO would officially declare a full-on pandemic. There are no hard and fast definitions for when other diseases, like Covid-19, can officially be called pandemic, but in 2009, that was the definition for a flu pandemic.

But unlike Covid-19, we knew about flu pandemics. Once it was declared, vaccine makers would activate the contracts for pandemic vaccine with the 15 countries that had them. Countries that had pandemic plans would activate them, closing schools and handing out the antiviral drugs we have for flu, depending on how virulent it was.

I started digging. The Mexican outbreak had started in early April, killed dozens of people including children, and spread widely over Easter—like Lunar New Year in China, the time Mexicans visit family. It started on a huge hog farm in Veracruz owned by US giant Smithfield Farms. The company protested that its pigs were all vaccinated and, anyway, had no flu symptoms. Well, yes, of course they didn't: vaccinated pigs don't show symptoms, but they can still carry and transmit flu.

The UN Food and Agriculture Organization announced it would mobilize its experts "to protect the pig sector from the novel H1N1 virus by confirming there is no direct link to pigs." Knowing the answer must have made it an easy investigation. The industry mostly seemed concerned about people calling it swine flu. Despite their efforts, most people still do. As far as I have been able to find out, the industry still publishes no monitoring data on flu in its herds.

The virologists I talked to were scared. This was H1N1, not just the same family as the H1N1 pandemic virus of 1918 but a direct descendant, passed down from the pigs we infected way back then. It also had some genes that came directly from bird flu, another unsettling similarity to its pandemic ancestor. So far, it seemed to cause fairly mild disease, although some people were dying, and they seemed unusually young. But then again, in the 1918 pandemic's first, spring wave, the disease was also quite mild. In the autumn, it turned lethal.

The WHO was supposed to declare a pandemic, once this new flu started spreading "in the community" outside North America— meaning, as it did for Covid-19, that there were people whose infection couldn't be traced to people or places already known to have it. But when Japan had a rash of cases, the WHO didn't act. Europe should have been next. Yet for some reason, cases were slow to materialize there.

On May 20th, I reported why. The European CDC had set rules requiring someone to have had exposure to the US, Mexico, or a known case to be tested, which basically precluded finding any community-acquired cases. Similar rules kept Wuhan from finding community-acquired cases of Covid-19 in January 2020, and then the UK, the US, and other countries from finding them that February.

The next week, two Greek students at universities in Edinburgh went to end-of-term parties, then developed fever and coughs on their way back to Greece. Their doctors defied the rule and tested. They had swine flu. The doctors complained that the rules were stopping Europe from finding local cases. I reported it in *New Scientist* on Friday, May 29th. The rules were changed Wednesday, June 3rd.

I honestly don't know if our article made a difference—the Greek doctors were the heroes. But after the pandemic was over, I got an unexpected gift from a flu guy at a health agency that shall be nameless: a commemorative swine flu T-shirt he'd had made for his staff, designed like a rock and roll tour shirt with the dates the virus arrived in different countries. It remains a treasured possession. On June 11th, with cases mounting in Europe, the WHO finally declared swine flu a pandemic.

On the day the WHO declared Covid-19 a pandemic in 2020, the agency complained about countries that still wouldn't test people who had no contact with a known case, China, or another early place that had the disease, even though it was clear the virus was spreading far more widely. It seems little has changed.

But in 2009, the WHO stressed that its declaration of a pandemic "was a reflection of the spread of the new H1N1 virus, not the severity of illness." It turned out Europe, Japan, and the US had been begging WHO not to declare a pandemic. All their pandemic plans were based on a worst-case scenario: a pandemic of a nasty bird flu that had been spreading around the world since 2004. Swine flu just didn't seem severe enough to warrant the upset. Declaring a pandemic would cause panic, they feared, for a virus that seemed to kill about as often as ordinary flu, albeit younger people.

Yet while governments were complaining that the pandemic virus was too mild to make a fuss over, vaccine makers had actually been counting on a mild first wave. That would give them time to make a vaccine in time for an autumn wave, which, if this was like the 1918 H1N1, could be dire. In the end, no vaccine was available until the autumn wave in North America was practically over anyway. There seemed to be a real mismatch between the pandemic plans, what we know flu does, and what we could realistically do about it.

Doctors I spoke to at the time said the new flu was mostly mild, except when it wasn't, whereupon it was horrendous—much like Covid-19. I remember a doctor in Winnipeg almost weeping on the phone as he described wards full of desperately ill young adults, many of them First Nations people who were especially at risk, needing ventilators and artificial breathing machines.

Scientists at Imperial College in London actually made some good come of this, by doing demanding clinical research "in the teeth of the pandemic," as one described it, and discovering a gene that predisposes some people to severe cases. It's amazing how many diseases of pandemic concern are like that: not so bad, except when they are. Being able to predict who is especially at risk would help protect those people, and maybe tell us exactly how these viruses cause benign or lethal disease, so we can design better treatments. The team is repeating the research with Covid-19.

However, swine flu pulled a real surprise that shows just how different flu is from Covid-19. People born before the H2N2 pandemic of 1957, when H1N1 still ruled the earth, were more immune than had been initially expected to the H1N1 virus of 2009. You have the strongest immunity to the first kind of flu you encounter as a child, for reasons no one yet understands. Before 1957, the only human flu around was an H1N1 descended directly from 1918. So was the pandemic virus, but immunologists initially thought the two were too different, and immunity to the old virus wouldn't protect people from the new one. They were wrong.

So the old people who normally die in droves from flu—and are currently dying of Covid-19—didn't in that pandemic, one reason it's considered "mild." Something similar happened in 1918, says Jeff Taubenberger of the US National Institutes of Health: elderly people born around 1850 were also relatively immune to the Spanish flu, possibly because a flu with similar surface proteins circulated then.

You hear stories that younger people died in 1918, but older people did not, because in both cases their immune reactions killed them, but younger people have "stronger" immune reactions. Nonsense. Older people had experienced that virus before, because flu varieties come and go with successive pandemics. That happened again in 2009. If H2 ever comes back, only people born between its arrival in 1957 and disappearance in 1968 will have much immunity. I hope that's good news for some of you.

Fortunately, the swine flu of 2009 never did mutate to become more severe, possibly because it was, well, swine flu—already adapted to mammals like us and not, like the 1918 virus, largely adapted to birds. After 2010, it settled down as a normal winter flu. It didn't even displace the H3N2 virus that had already been circulating, and the two now vie for dominance each winter, one or the other winning out in different places. Where H3N2 wins, more older people die.

Make no mistake: swine flu was not benign. Three times more children died than in a normal flu season. "I think it would be very misleading to describe that as mild," said Tom Frieden, then head of the US CDC. Estimates vary, but at least 200,000 people, and possibly as many as 600,000, died worldwide from swine flu in its first year, some 80 percent of them under age 65. Normally, the US CDC reports, 80 percent of people who die of flu are over 65.

All of this experience affected how we handled Covid-19, and will handle any future pandemic, flu or not. After the autumn wave, the WHO was bitterly attacked, amid widespread claims that responding to the 2009 pandemic—even when we had every reason to fear a replay of 1918—was an expensive overreaction. Some now even put a positive spin on that claim: in March, Zeng Guang, a top epidemiologist at China CDC, told the government newspaper *Global Times* that China "overreacted" in 2009, which "served as a public mobilization drill for all-round control and prevention in the face of a massive outbreak" like Covid-19.

But part of the attack on the WHO was led by the kind of people who have come to be called denialists: people who reject scientific information—even observable reality—that doesn't fit with claims that we are all victims of a giant conspiracy between big companies, corrupt governments, and (to them) shadowy scientists and international agencies. Swine flu was not really a pandemic, they claimed, even though it met every definition of one. Some of the accusers seemed disappointed it hadn't killed more.

The WHO only called a pandemic, the denialists alleged, so its cronies in the drug companies could make money selling pandemic drugs and vaccines—even though some vaccine companies actually took a loss on them, as a few countries asked for their money back. Individual scientists were accused of supporting the pandemic declaration because they were in the pay of these companies—a claim that did not withstand close examination, but was made easier because drug and vaccine companies, unsurprisingly, fund a lot of drug and vaccine research. If any of it had been remotely true, it would have made a great story for a journalist like me. It was frankly worse than untrue. It was poisonous claptrap, of the kind that has only gotten louder in the years since.

For example, when the Covid-19 pandemic started, I heard people I had thought were sensible claim that this was just another scam to make money selling vaccines. (Does any reader who has had the disease feel it was a scam? And, sorry, but what vaccines?)

Worse, though, in the years following the swine flu pandemic, the WHO seemed to become very gun-shy about calling pandemics. It stopped trying to set an official definition for pandemics at all, even of flu. In the early days of Covid-19, journalists kept asking: Is it a pandemic yet? WHO spokespersons got very annoyed. They asked, why do you care so much about that word?

I will stick my neck out here and respond: Because for years, the WHO has very correctly been warning us about the dangers of pandemics, especially of flu, which we know happen regularly. The word means something. The day after the WHO declared Covid-19 a pandemic, the world's press, much of which had been covering the story on inside pages, moved it to the front page. Some countries called their top-level emergency committees to discuss Covid-19 for the first time. There was an explosion of comment on social media. It made a big difference in how seriously people were taking this disease, and I think we could have used that a week or two earlier.

Perhaps the WHO is still smarting from being attacked for declaring a flu pandemic in 2009—even though it was a textbook flu pandemic—and wants to be very careful with the word. If so, the world's reaction to the swine flu pandemic did us all a disservice when Covid-19 arrived.

The WHO was also worried that governments would somehow conflate pandemic and flu. That's one reason flu matters so much to the story of Covid-19.

When Covid-19 hit, most governments with pandemic plans had based them around flu: many are actually entitled "influenza pandemic plan." Covid-19 is not flu, and that caused problems. Containment, where you isolate cases and trace and quarantine their contacts, was the WHO's main recommendation for Covid-19 early in the pandemic. But that is not possible with flu because the virus spreads faster than Covid-19, so it wasn't a part of these pandemic plans. The lesson here: plan, but be prepared for the unexpected.

To be fair, there was a very good reason governments had pandemic plans based on worst-case scenarios and flu. In 1997, 18 people in Hong Kong were infected with a bird flu called H5N1. Six died. Virologists were shocked: it was the first time anyone had seen a bird flu directly infect people, and the results were apparently lethal. Hong Kong killed all 1.4 million chickens, ducks, and geese in the territory to stamp it out.

H5N1 reappeared in 2001 in Hong Kong—which killed all its

poultry again—and 2002. In 2003, a family of four from Hong Kong contracted the virus while visiting mainland China. Two died. In January 2004, the virus ripped through poultry in Vietnam, and ten people had died of it when I ran into Ab Osterhaus looking shell-shocked at the Royal Society.

He had good reason. Overall, nearly two-thirds of people who have caught this virus have died. It hasn't been able to spread person-to-person, but Ab was worried it would learn to do that without becoming appreciably less lethal. If someone helping dispose of the millions of sick birds had human flu and got H5N1 as well, a recombinant might emerge with an H and/or an N no human had seen before—even the N1 protein was subtly different from its cousins on human H1N1 flu viruses. Add a few more bird genes, and it might not be much less lethal than the bird virus. The real nightmare might be if the bird virus itself could adapt to spread in people.

By late January, South Korea, Japan, and Cambodia had millions of birds sick with H5N1, and Thailand and Indonesia admitted that poultry die-offs they had been blaming on other diseases since the previous year were actually H5N1. Thailand had sick people. No one had ever seen bird flu over such a wide area. China reported a few dead birds just over the border from Vietnam and claimed it had only just got the virus.

Scientists I talked to didn't believe it. In 1999, a goose from mainland China carrying the same H5 as the 1997 virus turned up in Hong Kong, and the paper reporting it was entitled, "Continued Circulation in China of Highly Pathogenic Avian Influenza Viruses." In 2002, scientists at Hong Kong University reported finding a large variety of H5N1 viruses in chickens, which were probably "now widespread in the region [China]" and "justify renewed pandemic concern." So we had reason to think H5N1 was circulating in Chinese poultry.

It turned out that after Hong Kong slaughtered all its chickens in 1997, Chinese poultry producers selling to Hong Kong had started vaccinating their birds. That sounds like a good idea. But, American scientists told me, Mexican poultry producers had tried that too, and discovered that the bird flu virus could circulate at low levels in vaccinated chickens, without giving itself away by causing symptoms.

So I called the WHO official in charge of flu. I reached him on his cell phone as he sat on a bus heading up to the ski slopes—it was that time of year in Switzerland. He told me the WHO knew of virus samples from early 2003 that precisely matched the current outbreak—meaning this had been going on for a while. He wouldn't tell me which country they came from. On the 28th of January, I wrote that this outbreak had started a year ago and, based on what the scientists had been saying all along, probably in China—but poultry vaccination let the virus spread unseen.

The following day, China's vice minister for agriculture called a press conference in response. "It is purely a guess, a groundless guess," he fumed. "We have had strict surveillance." A spokesperson for the foreign ministry said the article was "completely inaccurate, without proof and moreover does not respect science." I started getting abusive emails from Chinese students. One accused me, presuming I was British, of complicity in the Opium War.

But the day after the press conference, Chinese officials confirmed that there were outbreaks of H5N1 in chickens in Hubei and

Hunan, just north of the outbreaks near the Vietnamese border. Two days later, there were "suspected" outbreaks in three more nearby provinces. The day after that, four more provinces had outbreaks, plus the huge western expanse of Xinjiang. Two days later, two more, this time northern provinces.

It was as if H5N1 was—improbably rapidly—moving across the country from its toehold near Vietnam. I have it on good authority that our report helped prompt the wave of candor. After all, if we could work it out, others could.

Sure enough, on February 2nd, *The Times* of London reported the following: "A large number of poultry markets in southern China have reported cases of the disease, and dozens of traders and butchers in contact with infected chickens have died." Chinese journalists were forbidden to report the deaths. There were no press conferences this time. Two months later, when I wrote about the further perils of chicken vaccination for *New Scientist*, Chinese state media reported it neutrally.

By then, scientists comparing H5N1 across East Asia had found it was all very closely related, but its surface proteins were changing fast. "We have a bucket of evolution going on," New Zealander Richard Webby, a leading flu virologist based in Memphis, told me. "This shows that H5 is circulating fairly widely somewhere, under some kind of unusual selective pressure."

Scientists who developed flu vaccines for poultry had warned about this in 2003. The vaccine might increase the risk of a flu pandemic in people, they feared, because the vaccinated birds spread the virus silently—and were a novel environment for a flu virus, so the viruses would probably evolve.

The fact that flu was in chickens at all wasn't normal. Flu evolved to live benignly in the guts of waterfowl—they poop it out, other ducks drink it, the virus continues. The virus needs the duck to keep shedding it for a while until, despite being diluted in pond water, it manages to reach another duck. The viruses that didn't make ducks sick therefore won the evolutionary race.

Chickens are a different matter. Most of the 19 billion chickens alive in the world at any time live in large henhouses. In such pastures of plenty, a bird flu virus left by a passing duck infects chickens, then after a while often develops a "highly pathogenic" mutation in the H protein that allows it to infect cells throughout the bird, not just the gut. The virus doesn't need to persist in its host and be shed for a while before it gets a rare chance to infect another host. Hosts are everywhere. The virus that wins is the one that replicates massively and then gets into the next chicken faster than the next virus. Chickens die in droves, but sometimes, it's adaptive for a disease to be more lethal.

In 2004, the chicken industry in East Asia, as in much of the world, had become large-scale and intensive, as growing prosperity boosted the demand for animal protein. And the H5N1 virus spreading across East Asia was a highly pathogenic strain.

Usually, these viruses kill their victims so fast they run out of them and burn out. But this H5N1 persisted because the vaccinated chickens didn't die. It did, however, have to contend with the chicken's novel immune system, which meant new pressures to evolve.

By 2006, Yi Guan of Shantou University had collected enough anal swabs from poultry across Southeast China to demonstrate that the virus had been circulating continuously there for a decade, virtually entirely within the poultry trade. There were growing concerns that it might adapt to spread readily between mammals, especially after high-profile tiger deaths in zoos. The tigers weren't the only mammals dying of the virus. "Javanese farmers even have a word for the cat disease," said Ab Osterhaus. In theory, every infected mammal was a chance for the virus to adapt to us.

Epidemiologists, including Ferguson's team at Imperial College London that did many early analyses of Covid-19, started making contingency plans for if that happened. Plan A was to watch for the first cluster of human cases and contain everyone exposed until the virus died out. If we failed to contain it—and some epidemiologists thought that was a long shot—then Plan B was to protect everyone with a vaccine or drugs. Sound familiar?

At the time, it seemed that governments did not understand that these were the only options on offer. Few had plans for deploying antiviral drugs or early surveillance to contain outbreaks—although to its credit, at this time China started developing its national computerized early warning system.

Plan B required developing a vaccine and, just as important, the ability to make enough vaccine and drugs for everyone. Vaccine companies have done some H5N1 vaccine development, but there was no way to make that much vaccine quickly enough. The threat is still there, and we still can't do this for flu, just as we can't for Covid-19—although Covid-19 has the excuse of being new. Flu is not.

In 2005, H5N1 spread beyond China, killing thousands of migrating birds at Qinghai Lake in the west of the country. Yi Guan found it was the virus from southeastern China. Senior Chinese officials attacked his science and made it illegal to collect animal disease samples.

I had been dubious of early reports that wild birds could carry the virus. But it had started to become clear that, although this H5N1 killed many kinds of birds—diving ducks, for instance, and swans—dabbling ducks seemed to carry it with few or no ill effects.

That was a problem, given the wandering habits of dabbling ducks. Male mallards migrate vast distances to the north in summer, nest on the tundra, and then fly back south. And they see the world: one year they may winter in Europe, the next Africa. Between the two seasons, they dabble in the same Siberian ponds as ducks from China. I ordered something I never imagined I'd need: an atlas of duck migration.

And it showed exactly where H5N1 went. Throughout 2006, the headlines came thick and fast, as H5N1 turned up in countries west of Qinghai, all descendants of the virus at the lake. There was panic in Britain and Bulgaria and Germany as H5N1 appeared, killing swans here, infecting ducks there. It turned up in northern Nigeria, just where my atlas marked a string of wetlands popular with mallards just in from Siberia. I got vicious emails from birdwatchers, terrified this would mean the persecution of birds. It didn't. But the virus itself has tragically harmed wild bird populations, no one knows how much.

H5N1 is still out there. There was a massive outbreak of H5N1 in chickens near Wuhan in February 2020. In March 2020, there were outbreaks of H5N1 and variants of it in poultry in Taiwan,

135

the Philippines, Afghanistan, China, India, North Korea, Philippines, and Vietnam, and in wild birds in Afghanistan, China, India, and Nepal. Among the virus's often-overlooked impacts is the cost, often to poor farmers, of destroying millions of birds to extinguish outbreaks. In Southeast Asia, that was estimated to have run into billions of dollars just by 2005.

Human deaths diminished after 2006, as people learned to avoid infected poultry. There were four deaths in Egypt and Indonesia in 2017, and then one young man died in Nepal in 2019. So far, 17 countries have reported 861 human cases. More than half died, a frightening rate.

The H5N1 strain is not the only problem. In 2013, an H7N9 turned up in live poultry markets in China and caused severe disease in people—but most of the viruses did not have the highly pathogenic mutation, so H7N9 didn't give itself away by killing poultry. Since 2013, H7N9 has infected 1,568 people in China and killed 616, 39 percent. Only four cases have been reported since October 2017. This may have been because China started widespread vaccination of chickens for it that year, reducing the amount of virus people encounte—but making its spread invisible.

A few human infections with H7N9 seemed to spread person-to-person after close contact. That was worrying. H5N1 has been circulating widely for 16 years now, yet it has never really done that. Virologists had long wanted to know if it was capable of evolving that capability. So Ron Fouchier—the Dutch scientist who first isolated MERS—primed an H5N1 virus with three mutations known to adapt bird flu to mammals. Two of them were discovered in the bird-derived proteins on the pandemic viruses in 1918, 1957, and 1968, so they had a history of enabling pandemics.

Next, he infected ferrets with the primed virus, caged them next to other ferrets—in a very-high-containment lab—and watched. Ferrets are the standard experimental animal for flu, as they get it much the same way humans do. The first set of ferrets spread it to a second, and so on for ten successive transmissions. This was it: mammalian transmissible H5N1. It turned out the virus acquired two further mutations in the ferrets that made it spread more easily. So H5N1 was only five mutations, in total, away from spreading in us.

None of the ferrets that caught the virus this way died. However, the team had discovered that, because our noses are different, you have to blow a virus into the ferret's windpipe to get an idea of how deadly it will be in people. When that was done, the transmissible virus killed all the ferrets. It had become fully contagious in mammals like us, without losing any of its deadliness.

Fouchier described this at a big flu meeting in Malta in 2011. As I listened to his talk, scribbling furiously in my notebook, I felt that weird mix of sensations you get on a story like this, both excited and terrified. This was big. People had started accusing scientists of hyping the threat of H5N1. Maybe it couldn't become transmissible in people, they said. *But this one did*, at least in mammals. I'll never forget how somber and serious Fouchier looked when I tracked him down at the coffee break and asked him about it. The other flu scientists I asked looked scared too.

Later, there was an almighty ruckus when the team submitted

the work for publication in *Science*, a leading scientific journal. The top biosecurity committee in the US tried to stop its publication, arguing that a bioterrorist might use the recipe to brew a deadly pandemic. As the work had partly been funded by a US agency, they had a say in publishing it.

Fouchier replied that we needed the work to understand what risk the virus posed, especially as it was now widespread in birds across Eurasia and Africa. He also claimed it wasn't such a scary virus after all—but I know what I saw on those faces in Malta. The work was ultimately published.

What did temper the scariness a bit is that not all of those three priming mutations had cropped up naturally in wild H5N1, although they had in other kinds of bird flu. Maybe for some reason H5N1 can't go transmissible on its own.

But here's the really scary bit: H7N9 already has three of the five mutations that made H5N1 transmit freely between Fouchier's ferrets. The fear is that if H7N9 infects the occasional mammal— which it does, humans—it could acquire the other mutations it needs while it's there, like H5N1 did in the ferrets. It may not need them, though: in 2017, some H7N9 acquired the highly pathogenic mutation in chickens, and Yoshihiro Kawaoka, a flu virologist in Wisconsin, discovered that those viruses would already spread between ferrets—and killed some of them, just from being inhaled, without the virus having to be blown into their windpipes. It's the first bird flu we've found that does that.

We don't know for sure if the mutations that made H5N1 transmissible in mammals also work for H7N9. The experiment hasn't been done. After the confrontation over publishing the H5N1 work, further work that might make nasty viruses nastier, called gain-of-function research, was banned or discouraged in the US and Europe. Anthony Fauci, the tough-minded head of the US National Institute for Allergy and Infectious Disease (NIAID), has become a popular hero in the US for calmly presenting the science of Covid-19 at televised presidential briefings. In 2012, he resolved the dispute over Fouchier's work by saying any future such experiments first had to be assessed for their risks and benefits by experts in the agency or it wouldn't fund them.

In 2017, gain-of-function experiments in theory were allowed to resume. There is certainly a real risk, and not just from bad actors deliberately making a bioterror germ—possibly worse is that other scientists with leaky labs might try to repeat the work. If such a virus escaped, it would no longer matter if it might have emerged in nature—we would have shot ourselves in our collective feet. I am personally inclined to believe Fouchier's lab is as safe as they get, as the Dutch inspectors are especially stringent, but I don't know about everyone else.

This issue goes beyond flu. Ralph Baric, who found that the coronaviruses Shi Zhengli found in bats in Yunnan could infect human respiratory tract cells, wanted to see what would be needed to make that virus more dangerous in people, like Fouchier and his ferrets. However, this would have been a gain of function, so he initially couldn't do the experiments. In 2019, NIAID funded the Eco-Health Alliance partly so that Shi Zhengli's lab, with its top-level containment facilities, could do that work with the bat viruses to see what changes to the external spike protein made them better at infecting human cells. It presumably vetted the risks and benefits first. The process is confidential, so all we know is that the vetting committee felt understanding the threat from these viruses outweighed the risk of escape. But the funding was cancelled after unsubstantiated allegations that Covid-19 escaped from the lab.

The scientists argue that we need to understand those viruses better, as nature is doing its own experiment. "Mother Nature is the ultimate bioterrorist" is the mantra I hear from virologists. The viruses Shi found naturally in bats were already plenty capable of infecting human cells, and a good example of what Mother Nature can do. After the funding was cancelled, EcoHealth Alliance released a statement noting that "international collaboration with countries where viruses emerge is absolutely vital to our own public health and national security here in the USA."

A flu pandemic is coming. That's what flu does. Maybe it will be fairly mild, like in 2009—but tell that to the people who lost loved ones then, and there were plenty. Maybe it will be H7N9, packing the same 40 percent death rate it has now. Maybe it will be a total surprise, something brewing on a giant hog farm or in a backyard flock of chickens swapping viruses with the wildlife. But it's coming.

Are we prepared? No. As we will see in the next chapter, we can't make flu vaccine fast enough, in large enough quantities for a flu pandemic. And although flu is the one virus for which we have effective antiviral drugs, it isn't clear we have enough of those either.

If we're not ready for the pandemic we can see coming, how can we be ready for the ones we don't?

## CHAPTER 6

# So What Do We Do About Disease?

> "The world needs to prepare for pandemics in the same serious way it prepares for war."
>
> —Bill Gates, Massachusetts Medical Society, 2018

The world was not prepared for Covid-19, and it is not prepared for pandemics generally. "In spite of all our 'alarmist' outcries in the past for better pandemic preparedness, we are now starting to prepare when the house is on fire," says Ab Osterhaus. What should we now be doing about that?

You'd have thought we wouldn't be lacking in pandemic plans: countries and experts have been talking about it ever since the world got spooked by H5N1 bird flu in 2004. Yet when Covid-19 arrived, there were disputes in many countries about whether to lockdown, how to do it, whether or not containment was possible, and when to lift restrictions. Instead of arguing about these things ahead of time, governments were vacillating just as medical staff ran out of ventilators and protective gear and the economic impact just of our efforts to slow the virus's spread caused mass unemployment, bankruptcies, poverty, even starvation. Very few governments seemed to have widely agreed plans for what to do when a pandemic struck; there was almost no coordination internationally, even, initially, within the European Union.

This shouldn't have been a surprise. Christopher Kirchhof, who led the US military's mission against the 2014 Ebola epidemic, described in March 2020 how a high-level analysis of the response to that epidemic concluded that with a trickier disease—one that, unlike Ebola, spread before it caused symptoms, like Covid-19—"the response system of the United States and the international response system would risk collapse."

The US tried to improve matters. It spent $1 billion on detection labs and preparedness plans in developing countries, as required by the International Health Regulations; stockpiled protective equipment and set up networks of hospitals in the US primed to respond to a pandemic; and created an office in the White House to plan and lead the response, the National Security Council Directorate for Global Health Security and Biodefense. All three, wrote Kirchhoff, were underfunded or shut down under the Trump administration. When the Covid-19 pandemic hit, the pandemic plan written by the Obama administration was largely ignored.

But while those political problems are unique to the US, lack of preparation and action was not. On March 11th, WHO director-general Tedros Ghebreyesus finally called Covid-19 a pandemic and said he was doing it because "we are deeply concerned both by the alarming levels of spread and severity, and by the alarming levels of inaction."

For weeks, the world, especially the rich West, seemed locked in a slow-motion train wreck, as though countries could not believe the oncoming storm was going to reach them and were paralyzed not knowing how to respond. There was a lot of denial: senior officials in North America and Europe were saying this might all still be contained in China when scientists suspected it was probably already worldwide—and, it turned out, those scientists were right. National plans that called unambiguously for certain responses when certain kinds of events happened should have triggered more decisive action earlier. Clearly, many countries did not have them.

Even where there were plans, and even if they were followed, they were mostly devised for flu, which as we have seen is different from Covid-19 in many ways. Containment doesn't work for fast-spreading flu, but as China showed, it works for Covid-19. The WHO delayed calling Covid-19 a pandemic partly because they feared countries would abandon containment and testing and rush straight to flu-inspired social distancing—and for some countries, it may have been right about that.

Many countries at least tried to plan for a flu pandemic. But when a milder-than-feared pandemic hit in 2009, some countries actually rolled back even that preparation. A Global Preparedness Monitoring Board (GPMB) co-chaired by Gro Harlem Brundtland, the WHO director-general during SARS, reported in 2019 that, "For too long, we have allowed a cycle of panic and neglect when it comes to pandemics: we ramp up efforts when there is a serious threat, then quickly forget about them when the threat subsides. It is well past time to act."

Yes, but what action do we need? There was some hope after the 2014 Ebola epidemic in West Africa almost spun out of control that the near-miss would jolt the world into doing more to prepare for major disease incidents. And it did trigger a few things that have been invaluable in dealing with Covid-19, such as the creation of CEPI, which organizes funding for pandemic vaccines, and the beefed-up emergency response capability at the WHO.

But we were still caught flat-footed. Work on drugs and vaccines for coronaviruses had been minimal even though we knew the risk. Some basic research had been done, and some start-up companies even had a few experimental vaccines, but nothing ready for prime time. Pandemic plans varied from country to country or state to state or didn't exist. A high-level UN panel warned in 2016 that the world was underestimating the risk of something less readily controllable than Ebola—such as a virulent respiratory pathogen—and that its ability to prepare, never mind respond, was "woefully insufficient." Covid-19 is a virulent respiratory pathogen, and they were right.

The "need for speed" the WHO warned of failed to register in all but a handful of places, as we have seen: South Korea, Taiwan, Singapore, Hong Kong, and a few others such as Vietnam and New Zealand. After Ebola, the UK set up a rapid response team that, it proudly said at the time, could investigate and respond to disease outbreaks anywhere in the world within 48 hours. But when Covid-19 arrived in the UK, the response was far slower. An initial plan for quarantine and contact tracing to achieve containment was let down by inadequate testing, then abandoned for a scientifically half-baked plan to allow most people to be exposed in order to develop "herd immunity." This was in turn abandoned when scientists explained how many deaths this would entail. It was replaced by social distancing, but the delay, plus weak enforcement, led by May 2020 to Europe's highest death rate.

Meanwhile limited protective equipment took a high toll of health care workers. The GPMB found that "the great majority of national health systems" couldn't handle the large influx of patients they would get with a severe, fast-spreading respiratory pathogen. There was no surge capacity in hospitals, they charged, or in key manufacturing, like making medical masks and gowns. That came true in many places with Covid-19.

At least the world's governments have now responded to the Covid-19 crisis by admitting there's a problem. That could be the real silver lining of this pandemic: there is no longer any dodging the fact that humanity is at risk from fast-spreading infectious disease and currently can apparently do little to prevent it or to respond effectively. On March 26th, the G20, the world's 20 richest governments, issued a statement promising "to strengthen national, regional, and global capacities to respond to potential infectious disease outbreaks by substantially increasing our epidemic preparedness spending."

And they did their homework on what the spending should cover. "We further commit to work together to increase research and development funding for vaccines and medicines, leverage digital technologies, and strengthen scientific international cooperation...rapid development, manufacturing and distribution of diagnostics, antiviral medicines, and vaccines, adhering to the objectives of efficacy, safety, equity, accessibility, and affordability. We ask the WHO...to assess gaps in pandemic preparedness and report to a joint meeting of Finance and Health Ministers in the coming months, with a view to establish a global initiative on pandemic preparedness and response."

And this time, no dropping the ball when the crisis is over, they promise. "This initiative will…act as a universal, efficient, sustained funding and coordination platform to accelerate the development and delivery of vaccines, diagnostics and treatments."

Sounds good, if they stick to their promises. It's a lot of very broad commitments though—how can they put it all into practice in an ultimately effective way? What exactly should they be working on, considering that, besides flu, we don't know what virus will be next—or even what kind of flu? What is a "substantial" increase in funding from governments reeling with the costs of the pandemic? The planning effort will at least be aided by the fact that now, with Covid-19, we can see clearly what we should have been doing for the past ten years. Let's look at the toolbox.

First, know your enemy. What germs should we target? And must we focus exclusively on responding to a disease once it emerges? Can we do more to stop it from emerging at all?

We looked earlier at the WHO's list of priority pathogens to get a feel for what may be out there. But not everyone thinks such lists are helpful. In 2018, the Johns Hopkins Center for Health Security warned that lists like the WHO's "stultify thinking on pandemic pathogens" by suggesting those are the only diseases we need to worry about. And some pathogens on the WHO list, they implied, were not truly global risks, but were put there to please regions where they are a problem. Instead, Hopkins called for keeping a close eye on the entire class of pathogens they suspect are most likely to cause real problems: respiratory RNA viruses. They mutate and evolve faster than any other pathogens and thus can jump species fast. And, contends Amesh Adalja at Hopkins, although we can stop gut pathogens with sewage management and infections like Ebola and HIV by being careful about body fluids, respiratory infections are harder to stop: no one can stop inhaling. Two years later, Covid-19, a respiratory RNA virus, proved their suspicions right.

Hopkins also called for more investigation into what pathogens are out there actually making people sick. Many people don't realize that most diagnoses doctors make are "syndromic"—pneumonia, meningitis, fever, and sepsis are terms that describe the disease process, not what is causing it. The actual pathogens behind these are often not even determined, as this is not needed for treatment. Instead, doctors use wide-spectrum antibiotics for bacteria, or with viruses—like Covid-19—they just try to keep the patient alive until the patient's antiviral immune response kicks in.

"Illuminating this biological dark matter," the Hopkins team argued, "would focus pathogen discovery efforts on established damage-causing microbes." Aggressively doing such diagnosis in a few sentinel locations, perhaps in zoonosis hotspots, might reveal the next big threat early as it starts jumping to people.

To do more specific diagnosis of syndromes, hospital labs need new kinds of diagnostic technologies that can distinguish a wide range of pathogens. This is why "diagnostic tools" is one of the things the G20 said they'd make sure we get. Fortunately, there has been a massive surge in these over the past decade, so we're at least on the right track.

Diagnostics manufacturers are now making automated "panels" of tests that can recognize the DNA or RNA of, say, a dozen respiratory or gut viruses, a huge improvement over old methods based on growing pathogens in culture to identify them, which is slow and insensitive. Making this kind of capability available more widely, including in countries that cannot afford it now but have hotspots of disease emergence, would illuminate a lot of pathogenic "dark matter" and give us a much stronger idea of exactly what pathogens we are contending with.

For now, the most widely used diagnostic panels on the market are designed for routine hospital practice and mostly look for the usual suspects that cause most human infection. In sentinel sites looking for surprises, it would be good to have something that can spot the unexpected and unknown. That might seem impossible—how do you design a test for the unknown? But one system, put on the market in Europe in 2014, could do just that.

The IRIDICA system was based on replicating the DNA or RNA from the pathogen in a sample and then putting it through a small mass spectrometer, which precisely determines its molecular weight down to the last atom. Using a database of known weights from different pathogens, the molecular weight can identify the species, and even whether bacteria are carrying an antibiotic-resistance gene. Or if it doesn't match any known species, says Rangarajan Sampath, chief scientist at FIND, a non-profit in Geneva that promotes diagnostics development, it can tell whether it is a hitherto unknown flu, or coronavirus, or a member of other virus families.

It was initially sponsored by DARPA, the US defense research agency that (DARPA people always mention) invented the internet. It was originally intended to scan for biological weapons—I heard about the prototype at a biodefense meeting in Stockholm in 1998. In 2009, an experimental prototype of the technology was the first to spot the new flu from Mexico in the US. It was set up to recognize not just flu but each of its eight RNA elements—and it spotted that this new virus had elements from bird, swine, and human flu viruses.

But doctors and regulatory agencies for medical technologies have been slow to warm to automated diagnostics. IRIDICA was finally put on the market in Europe in 2014 and was on track for approval in the US. Then, in 2017, the pharmaceutical giant Abbott, which owned IRIDICA, simply stopped making it. It had been a tough sell to hospitals with budgets stretched by government cuts and the growing health demands of an aging population. The real problem was that identifying the pathogens causing their patients' sepsis or pneumonia wasn't obviously cost-effective if there was often no pathogen-specific treatment to use as a result.

"It deeply saddens me," says Sampath. "There is still no viable alternative," he laments, especially for rapidly diagnosing pathogens causing sepsis, which is often lethal and where knowing the pathogen fast really can save a life. It would be invaluable in places like zoonosis hotspots, where real novelties can turn up, as it can rapidly rule out virtually all known pathogens and pinpoint an unknown pathogen's family.

There is a vicious circle in trying to promote more specific diagnosis. As we noted earlier, there's no point testing infections routinely if there isn't a specific treatment for the pathogen. But since we don't test, the Hopkins group argues, we don't know what pathogens we should be developing treatments for. Astonishingly, there are no specific treatments or vaccines for any respiratory RNA virus besides flu. At least we might soon have a few for Covid-19.

Moreover, there is global surveillance only for flu. Countries determine what varieties they have circulating and send samples to a global network of labs organized by the WHO, which is how we keep tabs on flu evolution and make new vaccines every year against what we think will circulate next. One perk for participating countries is that, in theory, they get access to any pandemic vaccines that result—although in an emergency, with vaccine-manufacturing countries tempted, despite international agreements, to hang onto whatever vaccine they make, it isn't certain the guarantee will work.

Hopkins wants surveillance extended to respiratory RNA viruses other than flu, with sampling from around the world, especially in suspected hotspots. They want it to include coronaviruses, Nipah and Hendra viruses, and enteroviruses, the most common family of viral infections in humans, which are mostly symptomless or mild—except a few, one of which is polio. They even want to watch rhinovirus, the only more frequent cause of common colds than the four mild coronaviruses that circulated in people before Covid-19.

Because even common colds can go bad. A cold is a syndrome, not a specific germ: at least 200 viruses cause them. In 2005, a novel adenovirus, Ad14, appeared on US military bases after they stopped vaccinating for adenoviruses, which are common causes of colds among recruits. This one caused severe pneumonia in 140 known people in the US, many of them young and healthy, and there were probably many more cases that were not tested. Ten of the 140 died. By 2008, most people were exposed and immune, and it became just another winter cold virus—but viruses don't always settle down and behave like that. And we don't know about most of them.

Some think we should find our plagues before they find us. Of course, we should try and spot new diseases after they emerge in people, so we can shut them down fast. But Peter Daszak observes that with both novel disease outbreaks and their economic impacts increasing fast, it would make sense to deal with the underlying drivers—changes in human ecology and human–animal interactions—to stop emergence from happening at all.

To enable that, the proposed Global Virome Project wants to genetically sequence and map the estimated half-million viruses in animals and birds that belong to families of viruses that we know can infect humans. It would cost $3.7 billion over the next ten years, says project leader Dennis Carroll, who also launched the PREDICT program that helped discover the bat viruses in Yunnan. He contrasts that to the trillions the Covid-19 pandemic will cost. Knowing where the potentially dangerous viruses are, he says, will help us focus preventive efforts, such as reducing interactions between people and the species or the places that we know have worrying germs.

But critics object that although such a survey would be great science, it isn't much good for preventing the next pandemic unless we also know what the viruses can do—and then do something about them. "These efforts will not necessarily translate into better pandemic preparedness, given the sheer numbers of viruses that will be catalogued without a clear means of prioritizing them, [and] the fact that most identified viruses will pose little to no threat to humans," says the Hopkins team.

"No amount of DNA sequencing can tell us when or where the next virus outbreak will appear," agreed Andrew Rambaut and colleagues in a critique of the idea in 2018. The 2014 Ebola epidemic was, at that point, the most-sequenced viral outbreak of all time—and that didn't stop another in the Democratic Republic of the Congo in 2018. Indeed, by 2013, virologists had sequenced and reported SARS-like bat viruses and warned of their pandemic potential. "That prediction didn't stop Covid," says Adalja. "People think the risks of these things are hypothetical." As we all know now, that one wasn't.

Echoing the Hopkins group, Rambaut and colleagues say it would be money better spent to do disease surveillance in people, to spot new infections as they emerge, using sequencing to spot the viruses and serology—testing blood for antibodies—to see what infections people have had before.

And this, they say, would be best done by a global network of trained, local researchers. There Carroll agrees, although he also wants this network to "monitor, respond and prevent viral spillover while they are still evolving in animal populations." (Case in point: bat viruses.) Even scientists who disagree on where in the process of viral emergence we need to look agree on one thing: we need more people, everywhere, looking, preferably in their own backyards.

## So What Do We Do About Disease?

\* \* \*

We do have some eyes on the world already. The Canadian system that first spotted SARS is still watching global online chatter for mentions of disease and sends WHO about 3,000 "signals" per month, online mentions of things that might bear watching. The WHO follows up on about 300 of these, and investigates 30 in more detail—on average, one a day.

But veterans of the international health scene, like David Heymann, and Seth Berkeley, head of GAVI, an agency that helps poor countries buy vaccines, say more countries should do their own monitoring and share the results. Eavesdropping on global online chatter is an interesting way to keep tabs on things, but to seriously monitor disease, local public health people with a good idea of local diseases, and the ability to investigate, would be optimal.

This is not a new idea—it just hasn't been acted on. The International Health Regulations (IHR) were originally a binding international treaty based on earlier rules, drawn up in the 19th century, that required countries to notify each other about a few diseases—cholera, plague, yellow fever, and, before it was eradicated, smallpox—that posed an international risk through shipping.

After SARS, the IHR treaty was revised. The 2005 version obliges countries to save lives and jobs endangered by the international spread of any disease. They are supposed to coordinate their monitoring and response to disease with each other, and rich countries are meant to help poor ones do sufficient surveillance to spot anything dangerous.

They have done a bit, but not enough. When Ebola broke out in

West Africa in 2014, the first thing that failed was surveillance. The outbreak began in Guinea in December 2013, but it wasn't identified as Ebola until March, by which time it had spread widely. After that, the response failed: it was August, and the virus was out of control in two cities, before the WHO declared an emergency.

The WHO was criticized for the second delay, which happened partly due to organizational rigidities it has since tried to fix. But the fundamental problem was the first delay, and that was failed surveillance. The IHR requires countries to tell the WHO about any outbreak that is serious, unusual, or could trigger international travel or trade restrictions. That applied to Covid-19, and China did tell the WHO about it, but there are no provisions allowing the WHO to inspect the situation on the ground to see if the declaration is true—for instance, whether the infection really wasn't spreading between people.

Most concerns, however, have focused on countries much poorer than China that don't have the capacity to detect and diagnose a sudden cluster of infections and tell the WHO about them. Many such countries are in exactly the tropical or subtropical hotspots for disease emergence that need closest watching—it was to investigate such alerts that the UK developed its 48-hour response team. When the IHR treaty was updated in 2005, it required all countries to put their own surveillance capability in place by 2014. It then extended the deadline to 2016. Did countries make it?

"They didn't," says David Heymann, who steered the negotiations that revised the IHR in 2005. "Rich countries have been more interested in funding international response capabilities," like the WHO's new emergency unit. "There's been much less help for poor countries to take charge of their own pathogens surveillance." It's almost as if rich countries are interested in riding to the rescue in emergencies, but not in preventing the disease emergence that causes emergencies in the first place. In fact, surveillance and response need to go hand in hand.

An assessment in 2019 by an international collaboration called the Global Health Security Alliance, aimed at measuring and enabling countries' adherence to the IHR, found that "no country is fully prepared for epidemics or pandemics," whether it was rich or poor. Countries were judged on, among other things, whether they could prevent the emergence of pathogens, detect—and report— epidemics "of potential international concern," respond to them, treat the sick, and protect health workers.

The results were abysmal. Only 19 percent of countries got marks over 80 percent for detection and reporting capabilities, whereas fewer than 5 percent got top marks for their ability to rapidly respond to and mitigate an epidemic. The overall average in all categories was 40 percent, rich and poor together, but even rich countries on their own only got an average score of 52 percent.

Pandemic preparedness surged a bit in rich countries after the anthrax attacks of 2001 in the US and after the threat of H5N1 bird flu in 2004. But when the 2009 flu pandemic wasn't apocalyptic, preparedness fell out of favor. Some countries let stockpiles of antiviral drugs for flu lapse, and as far as I have been able to find out, no one renewed preorders for pandemic flu vaccines.

The UK government held a simulation drill for a flu pandemic in 2016, Exercise Cygnus. Its results have never been published, but participants have let slip that it showed that health services, and even morgues, would be overwhelmed. That happened with Covid-19, so we can only assume the warning was not adequately acted upon. Activists are launching legal action to compel the release of the results. A similar flu simulation exercise in the US in 2019, called Crimson Contagion, also not released until leaked by the *New York Times* in March 2020, revealed a confused, underprepared pandemic response that foreshadowed what happened with Covid-19.

You would think if any organization was prepared for epidemics and pandemics, it would be the WHO, but its slow response to Ebola in 2014 proved otherwise. That was partly due to undue deference to its own African regional office and local governments, which tried to downplay the outbreak when it still seemed small, and partly, I heard at the time, to a rigid hierarchy that kept its epidemic experts in the field from alerting the leadership to what they knew was coming—that difficult point in an epidemic when it still looks trivial, but isn't. But it was also because the agency simply couldn't raise the cash for an emergency response fast enough.

But the WHO adapted fast. "During Ebola, we needed to do new things, like recruiting 2,000 people to isolate Ebola cases and their contacts in remote areas," says the WHO's Bruce Aylward, who ramped up the agency's Ebola response in September 2014. The WHO had to bring in new kinds of expertise from disaster response agencies. The learning curve was steep. "It's like asking a penguin to fly," he told me later. "You throw it off a cliff and I'm amazed how well the damn thing flew."

The WHO emergency response team that grew out of that experience has led the international response to Covid-19: the penguin has grown real wings. But it is supported by voluntary funding from member states, never completely reliable, and repeatedly imperiled, including at the height of the pandemic in April 2020, by threats from the US to pull funding. Yet to prevent future pandemics, the world needs reliable surveillance and response, and those need to be coordinated globally: the job of the WHO. It will need more funding, more reliably, to do that.

Moreover, with more surveillance, it will need to be able to predict, far better than we can now, which outbreaks pose a threat and require a response. Covid-19 was not the first unexplained pneumonia in China—earlier we saw there were instances reported on ProMED in previous years that seemed to go no further. Which outbreaks have legs, and what might tip us off?

There is hope that "big data" might help, with everything from Google searches for "flu" to hospitals' anonymized, electronic medical records crunched on a large scale to suggest when something serious might be emerging. Ultimately, more research into how viruses damage us, and what makes some worse than others, should start telling us how to spot the dangerous ones. Systems like China's reporting network, designed to reveal an unexpected cluster of a syndrome in a region before it may be obvious on the ground, might also help here.

In March, in a report for the think tank The American Enterprise Institute, scientists called for a permanent national infectious disease forecasting center in the US, to "function similarly to the National Weather Service" and provide "decision support" for public health, including what responses are warranted by what kinds of events. The researchers warn, however, that predictions in any complex system are no small undertaking. Weather is a fitting analogy, as the US National Weather Service spends $1 billion a year collecting massive banks of weather data to turn into forecasts. The US CDC spends a quarter of that on analogous data for public health—local incidences of disease, age-specific death rates, vaccination rates—and has no budget for forecasting, as that has never before been realistic in public health.

Of course, we already have one proven, crowd-sourced, worldwide, and battle-hardened disease surveillance system: ProMED. It also runs something called EpiCore aimed at getting around nations' reluctance, or inability, to report disease. Medical and veterinary workers who know some field epidemiology sign up for it, and then if ProMED hears a worrying rumor, it can ask them, privately, to check it out using a web platform guaranteeing privacy. If you meet the criteria for membership, you might consider joining.

Astonishingly, however, ProMED is supported by a few grants and voluntary donations and can barely cover its costs. Could we maybe find it some independent funding? It is an embarrassment that such a mainstay of the global response to infectious disease has to regularly ask for donations. In April 2020, as Covid-19 raged, we all got emails signed by Marjorie Pollack pleading for readers to maybe pony up $25.

Finally, if we're going to take surveillance more seriously, there is one more area besides zoonotic hotspots we need to watch: labs. As we saw in the chapter on flu, research funding agencies in the US and Europe are currently reluctant to permit experiments that, on purpose or not, make pathogens more dangerous, in efforts to find out how much threat they may pose. The scientists argue, though, that we need to know whether some viruses can actually become more dangerous, and if so what mutations to watch for. The US National Institute of Allergy and Infectious Diseases decided last year to renew funding for studies of coronaviruses at the Wuhan Institute of Virology for just that reason.

The dilemma is always that doing the experiments without very tight containment might create the disaster we are trying to avoid. We can do research like that safely; we have been doing it for years. But I would argue that along with more funding for research, we must also fund better containment in labs and more oversight to make sure researchers are working safely—and on things that truly warrant the risk.

I've heard about too many cavalier experiments, or planned experiments, with viruses over the years. There was a collective sharp intake of breath from the virology world in 2001 after a colleague on *New Scientist* broke the news that a lab in Australia had completely inadvertently created an extra virulent strain of mousepox, a rodent infection related to smallpox, by giving the virus a gene for what they thought was an innocuous immune-modulating substance. Then, in 2003, at a meeting in Geneva, I heard an American scientist describe plans to make an even deadlier version of that virus in a species of pox that humans can theoretically catch, although he was hoping the effect wouldn't manifest in people.

The late D.A. Henderson, who led the eradication of smallpox, was still alive then, and he was sitting next to me in the lecture hall, getting perceptibly angrier and angrier. Several other scientists in the room were also looking uneasy. When one asked what the researcher hoped to learn from the experiment to warrant such a risk, a voice in the back said, "Nine-eleven." Apparently, we had to do this because terrorists might. I don't know if those experiments happened.

One thing we can say about the virus that causes Covid-19, though, is that it was not made in a lab. In February, once they had had a chance to look carefully at the virus, Kristian Andersen of the Scripps Research Institute in La Jolla, California, and colleagues reported that, basically, virologists simply wouldn't have known enough to make a virus like this.

Proteins are strings of hundreds of a smaller kind of molecule called amino acids. The types and order of these amino acids determine the structure of the protein, which in turn determines what it can do—proteins are basically the tiny machines that do most of the processes of life. One spot on the Covid-19 virus's big external spike protein fits a spot on the ACE2 protein on human cells and binds to it so the virus can infect the cell. The binding site on the virus is a string of amino acids, Andersen admitted, that we wouldn't have predicted could bind to human ACE2. No virologist trying to build an artificial binding site would have chosen those. But it turns out they work just fine. The virus also has some completely novel mutations that would have been unlikely to arise in a lab just trying to study a wild virus and a viral "backbone" unlike anything used to study coronaviruses in the lab.

In March, an unprecedented statement by 27 of the biggest names in emerging disease ran in the top medical journal *The Lancet*. "We stand together to strongly condemn conspiracy theories suggesting Covid-19 does not have a natural origin," it said. Scientists from many countries had studied it, "and they overwhelmingly conclude that this coronavirus originated in wildlife." Calling the efforts of China's scientific community to deal with the outbreak and share their results "remarkable," they concluded, "We want you, the science and health professionals of China, to know that we stand with you in your fight against this virus....Stand with our colleagues on the frontline!"

That wouldn't be a bad rallying cry to carry forward into the post-Covid-19 world. There's no point doing all this planning, diagnosis, and surveillance without organizing a global response to whatever we find out is happening. For all of us, now, "disease anywhere is disease everywhere" is no longer a slogan that sounds like it's for some telethon, but a lived reality.

But as diseases have emerged, there has been an absence of globally joined-up thinking. Zika took the Americas by surprise in 2015, even though it had already moved eastward into the Pacific from Asia, causing more severe disease than it ever had before, and even though chikungunya, another mosquito-borne virus from Africa, made exactly the same trip in 2013.

When virologists in Wuhan and North Carolina found bat coronaviruses that made mice sick and easily infected human cells in 2013, they practically shouted about the pandemic concerns this raised. Scientists also did that in 2004 about the family of swine flu viruses that indeed went pandemic in 2009. Perhaps it's easy to spot that in retrospect. But both times, nothing much seemed to be done.

The One Health Platform is an organization of scientists, started by virologist Ab Osterhaus and his colleagues, that tries to bring together a wide range of researchers in human, animal, and environmental health and people from governments and international organizations to take a wider look at global health security. It is as good a place as any to start talking about how the world should build operational, day-to-day activities that really address the threat of pandemics—not just spotting novel pathogens, but doing something about them.

In a paper ahead of the World One Health Congress in 2020, the organizers say that in "peace time," between pandemics, we need the following: surveillance and diagnosis of disease syndromes in humans and animals; the identification of new pathogens; development of diagnostics and mechanisms for distributing them; research into how new infections cause disease; drugs; vaccines; and communication between scientists, governments, and the public. That last one especially tends too often to be forgotten.

But that's a true scientist's list—mostly about finding out, not about taking action. Action, after all, is the job of governments, not scientists. So we might add to that list an authoritative, international capability for deciding when all that investigation has in fact revealed a potential threat that requires a response—drug and vaccine development, active monitoring—and making the response happen. Now that it is wartime, maybe governments will start paying attention—and paying for that kind of preparedness. It has to be someone's job to organize the response to the warnings we will turn up as we redouble our watch on emerging pathogens.

So, about those drugs and vaccines. First, know your enemy; then, choose your weapons.

As of May 2020, CEPI was organizing research and trials on nine different kinds of candidate Covid-19 vaccines, and the WHO was helping organize trials of existing antiviral drugs to see if they worked on the pandemic virus. This is all essential, and we must be capable of such organization in wartime. But we need to look forward too, at least after the first rush of R&D for Covid-19 has started bearing fruit, or we will always be playing a deadly game of catch-up.

The one vaccine that desperately needs funding is one CEPI doesn't do, but was a priority the GPMB highlighted last year: a universal flu vaccine. You have heard this already, but it cannot be said too often: a flu pandemic is inevitable. A pandemic flu carries novel surface proteins to which many humans have little or no immunity. By definition, we cannot make a vaccine to a pandemic strain in advance, as we have no idea what surface proteins the next one will have: there are endless variations, and immunity to one does not give you immunity to the others.

We could cook up a vaccine after the pandemic strain emerges when we know what it looks like—in fact, this is the current plan, as it is pretty much all we can do. But doing that cannot protect enough people fast enough. The quantities aren't too constraining: the world can make 1.5 billion doses of winter flu vaccines per year, which means that, in theory, it can make 6.4 billion doses of a pandemic vaccine. There are more people in the world, but vaccine experts have told me we are unlikely to reach everyone even if we make more.

Why the difference, though, between ordinary and pandemic vaccine-making capacity? In most seasonal flu vaccines, one dose contains 15 micrograms of the H protein from each of the three strains of flu circulating every winter: the H3N2, the H1N1 left over from 2009, and the dominant strain of influenza B. Some vaccines have four strains, with an extra B. In theory, since a pandemic vaccine is targeted at just the one strain, a pandemic vaccine only needs 15 micrograms of that specific H. So when vaccine production lines switch from seasonal to pandemic vaccine, they are capable in theory of churning out enough flu virus for three to four times more individual shots than normal.

However, we might turn out to need far more of the pandemic virus's H protein to goad our immune systems into mounting an immune response, in which case we will be able to make far fewer doses of vaccine. That happened with some experimental vaccines for H5N1 bird flu. Or we may need two doses each of the vaccine a month apart to be protected, as younger people who had never encountered the H on the 2009 pandemic virus did. That will take time to administer, and by this point, we will not have enough for everyone.

Or, the outlook could be sunnier. We might be able to use an immune-stimulating chemical called an adjuvant to make small doses go farther: several good ones have recently been developed, one of which is being incorporated into candidate Covid-19 vaccines. Vaccine researchers have also studied the possibility of making doses go farther by using tiny micro-needles to inject that flu protein into our skin, instead of deep into a muscle as we do now. Skin is crawling with immune cells, that can make the most of a tiny amount of vaccine.

At least we can now make four times more standard flu vaccine than we could in 2006. One reason is because, as concerns over bird flu mounted back then, poor countries worried about their access to vaccines in a pandemic. So in 2006, the WHO launched a campaign to increase vaccine manufacturing capacity and put it in more poor countries.

But all of these plants use the standard process for making flu vaccine, growing flu virus in eggs, which takes six months to produce enough—and that's if the vaccine virus grows well. In the 2009 pandemic, there was no vaccine available until the end of the autumn wave, partly because the virus initially grew slowly. If that H1N1 had turned extra virulent in the autumn wave like its forebear in 1918, the fact that vaccines came too late would have been disastrous.

So despite all this effort to make more standard vaccine, we still probably can't produce enough, fast enough, to save a lot of people if a really lethal flu hits. There are a few proposals for growing made-to-measure flu vaccines faster, for example by producing flu proteins from the latest strain in plants.

But the Holy Grail of flu vaccination is a universal flu vaccine. Scientists have been working on this for some 20 years.

In theory, we could use bits of the flu virus that do not change, either year to year or between families of the virus, to immunize ourselves to all flu once and for all. Our immune systems mostly ignore these "constant" regions of the flu virus, seduced into making more antibodies against the big, obvious H protein—which is why flu viruses have it. The hope is that if we were more strongly immunized against these constant regions, our immune systems would attack any flu virus we encounter.

We could develop and test it, and get people immunized, before a pandemic starts, even though we don't know what the exact virus will be. We could even stockpile it for people who hadn't already been vaccinated when the emergency hit. Several candidate vaccines have passed safety tests and seem to induce the right immune reactions.

However, only one, made by the Israeli firm BiondVax and containing nine unvarying stretches of protein from flu, has so far managed to get funding for the expensive, large-scale trial needed to see if it works. That trial is still underway. It has been hard for companies developing these vaccines to find enough funding, for the usual reason: it would not be profitable for a company to make such a vaccine, as people would need only one shot or a few shots during the course of their lives. But no one other than the big companies can fund and organize such large-scale trials. The joke for years has been that a universal flu vaccine is always five years away.

The GPMB called on governments to set a time line, by September 2020, for developing a universal flu vaccine. It didn't hazard a guess as to what that time line would be. But there has been a steady buzz of low-key research for years, and a universal flu vaccine does seem almost in reach. How soon the world will buckle down on anything but Covid-19, though, is an open question.

It seems like all we need to finish the job is an intensive burst of coordinated research, large-scale trials of the best universal vaccine options, and money to build manufacturing plants for the winners. Covid-19 is showing we can do all of this when we need to. The start-up companies that develop the vaccines can't do it. If there was ever a case for public spending on a public good, instead of leaving things to a market that simply cannot do this job, universal flu vaccine is it.

In fact, that too could be a silver lining to the dark cloud of Covid-19. Public-private partnerships have been starting to work on needed but non-profitable medical technologies like medicines and vaccines ever since big philanthropies like Gates—actually, ever since Gates—got involved in R&D for the diseases of poor countries in the 2000s. Now that might accelerate.

The press has been awash with commentaries claiming big government is back, as only governments can rescue the various industries that are going under during lockdowns and provide emergency income for people who have lost their jobs to social distancing. A lot may depend on how governments decide to handle the debt that will result, but many voters may prefer their tax money to go to better medicines and pandemic preparation than to some kinds of bailouts.

It is accepted that governments invest in public goods—roads, schools—to provide the infrastructure that allows private enterprise to flourish, at least in theory. Market failure means we have no universal flu vaccine or effective antiviral drugs, and we're losing our antibacterial drugs to resistant bacteria—we'll get to that in a moment. If a government really wants to support its industry, there's a case to be made for keeping its workers and consumers alive.

Of course, it's not only flu that should worry us. Another useful thing to have, if it is possible, would be a vaccine platform we can use for any virus that emerges: the WHO list's Disease X. One plan is to have an already tested, safe vaccine technology we can customize with a bit of the new virus, so it can be deployed with minimal further testing. There's precedent here.

Two Ebola vaccines were developed amid the burst of anxiety and funding that followed the 2001 anthrax attacks in the US, when it was feared Ebola might be used as a bioweapon. Work on them petered out as that burst of anxiety did, along with the funding.

Also, with no sizable Ebola outbreak, the vaccines couldn't be tested.

Then Ebola struck West Africa in 2014, and big vaccine companies, to their credit, stepped up and organized trials. (Similar trials were organized for antiviral drugs, some now being used for Covid-19.) One of the vaccines, originally developed by the Public Health Agency of Canada, was 95 to 100 percent effective at stopping Ebola in contacts of cases. It is now called Ervebo, the first Ebola vaccine ever put on the market. Another vaccine was tested in the 2018 Ebola epidemic in the Democratic Republic of the Congo.

The point for future pandemics is that both vaccines consist of benign viruses carrying an Ebola protein. The protein induces immune reactions specific to Ebola, while the virus carrying it—a "vaccine platform"—gets the immune system's attention. Similar vaccines equipped with Covid-19 protein are now in development.

The hope eventually is to have just such a tested vaccine platform ready so we can just drop in a new protein from any surprise virus that emerges and have a vaccine quickly. CEPI wants to develop a safe, all-purpose system like this, so it can have a vaccine ready for tests in people within 16 weeks of a new pathogen being detected. That is not possible yet: the Covid-19 vaccine candidates that use a similar platform to the Ebola vaccines won't be through trials until the end of 2020. But those tests might bring an all-purpose vaccine platform closer.

There's a further complication, though. We can design and test as many vaccines as we like for emerging diseases and for pandemic flu, but it isn't clear where we would make them in the large quantities required. There will never be enough demand for ordinary seasonal flu vaccine, industry insiders admit, to warrant enough manufacturing capacity for all the vaccine that would be needed in a flu pandemic, for example. And how do we become equipped to make vaccines to a threat that has not yet materialized? Bill Gates is funding the construction of seven different kinds of vaccine manufacturing plants while candidate Covid-19 vaccines are tested, so the one or two that work best can be mass-produced immediately—even if the other plants must be discarded. That's how serious this problem is.

We can brew up enough vaccine in labs to do safety tests in humans. But what if we have a Nipah vaccine we know is safe and want to use it in Bangladeshi villages where people are dying of Nipah to see if it saves lives? We'd need more than a small lab can make.

We can't build a vaccine factory just for that vaccine if we don't yet know it works. There is little spare vaccine-making capacity. If the 2014 Ebola outbreak had needed more vaccine for trials than manufacturers managed to squeeze out with the little spare capacity they could drum up, production lines making important childhood vaccines would have had to be switched over. Fortunately, they didn't need it.

We might build vaccine factories just to have such spare capacity. But they aren't easy to keep in reserve, say experts at the WHO. You can't build one just for emergencies: it has to be working to keep its staff and processes up to scratch. In March, the American Enterprise Institute called for a dedicated program to develop "flexible platforms" for producing drugs and vaccines for a new pathogen "in months not years," including "flexible manufacturing capacity to scale up production to a global level in an emergency."

We will also need to find ways to distribute the drugs and vaccines we create so all parts of the world have equitable access— a point made strongly by the international vaccine agency GAVI for Covid-19. CEPI agrees: "The challenge we face is not only one of R&D but one of manufacturing at scale, and equitable access." On April 24th 2020, governments, the WHO, and philanthropies launched an Access to COVID-19 Tools Accelerator, a funding program just for that. With luck, that could establish a precedent for other public-good drugs and vaccines in the future. With less luck, international jostling for the rights to still-experimental Covid-19 vaccines will turn into a hostile global competition for vaccine access and ownership.

That would be tragic. Equitable access is not only the ethically correct thing to do, but it is also simple self-interest. Let's say we develop a good vaccine and one part of the world vaccinates its people, but another part cannot, and the disease continues to spread there. The diseased part of the world will continue to constantly pump out virus, which can reach anyone susceptible in the vaccinated part. Meanwhile, viruses evolve, unpredictably, as long as infections continue, and soon the epidemic virus might well be one the vaccines weren't made for. We truly are all in this together.

There is a more speculative area of therapeutics I should mention. Covid-19, like SARS, seems to kill by triggering out of control inflammation. Normally, inflammation is a general activation of the immune system that gets rid of infections—but it can get out of hand. The reason older people and people with underlying conditions such as diabetes, high blood pressure, and even obesity have a higher rate of severe disease and death with Covid-19—and flu— is because all those conditions, including aging, involve chronic inflammation. The virus triggers more inflammation on top of that, and things somehow get out of control.

Inflammation is fiendishly complicated, so it's difficult to tinker with, but some drug developers are now looking at ways to tackle excessive inflammatory responses themselves as a way to limit both the chronic underlying conditions and the impacts of infectious diseases, and maybe even some aspects of aging. A space to watch.

GPMB also called for work on broad-spectrum antivirals, analogous to the broad-spectrum antibiotics that kill a wide variety of bacteria. Such drugs can in theory be used to knock down any unexpected virus that emerges. But wide-spectrum antibiotics provide a cautionary tale: because they kill many kinds of bacteria, they also promote widespread antibiotic resistance. Viruses can develop resistance too. We have two families of antiviral drugs for flu, and genes for resistance have emerged for both of them. One kind is in the troublesome H5N1 bird flu we have studied.

Thankfully, the antiviral Tamiflu still works against most flu, and it is stockpiled in some countries in case of a flu pandemic. But this illustrates another kind of threat. There has been a denialist crusade against the drug and the pandemic stockpiles, based on claims that the manufacturer's drug trials show it doesn't do much against ordinary winter flu. One critic told a British parliamentary committee there was no evidence that Tamiflu was better than "a stiff whisky."

In fact, there is plenty of evidence. The drug is stockpiled for pandemics, not ordinary winter flu, and it is used in a pandemic to stop people from dying of severe viral pneumonia, similar to what Covid-19 causes. The drug trials were aimed at determining whether the drug affected ordinary flu, not severe pandemic flu. Jonathan Van-Tam of the University of Nottingham found that among 168,000 people with flu severe enough to need hospitalization in the 2009 pandemic, people who received Tamiflu within two days of falling ill were half as likely to die—a significant effect. We would love to have a drug like that for Covid-19. Yet the crusade against the flu drug continues, with a lawsuit launched against the manufacturer, Roche, in the US in January 2020 for allegedly "bilking" the US government of the money it paid for its stockpile.

This seems like a good time to address the elephant in the room whenever we talk about any aspect of our future health, including the risk of pandemic viruses: antibiotics, the drugs that kill bacterial infections. No one really expects a bacterial disease to stage a pandemic, although I am increasingly reluctant to rule out anything involving the living world. But bacteria aren't as rapidly evolving or communicable as many viruses—there's a reason we call something speeding across the internet "viral."

Yet antibiotics are crucial in a viral pandemic like Covid-19. In February, Chinese doctors reported that some 94 percent of patients with Covid-19 severe enough to be hospitalized were being treated with antibiotics because doctors feared that, as with severe flu, bacteria would take advantage and run rampant in their lungs. Doctors in the US and elsewhere are also using antibiotics with Covid-19 patients who need ventilators, which increase the risk of bacterial infection.

And antibiotics would assuredly be needed in a flu pandemic. In 1918, a third to half of the millions who died are thought to have been killed not by direct viral pneumonia, but by the bacterial pneumonia that often follows flu. Historians often reassure readers that 1918 could never happen again in our modern world because now we have antibiotics.

This easy assumption that we will always have effective antibiotics always makes me shudder. A growing number of such infections already resist antibiotics, and we get more resistant bacteria the more antibiotics we use. The massive use and misuse of antibiotics to desperately treat bacterial complications in a viral pandemic could very possibly accelerate the process. There are fears that it is happening now.

You have probably heard about antibiotic resistance. Many antibiotics come from microscopic fungi in soil, which use them in their constant war with soil bacteria. Accordingly, the bacteria have developed genes for proteins that block or demolish the fungal antibiotics. And bacteria share genes like foodies share sourdough recipes. Probably more.

If you expose bacteria to an antibiotic, some may well have a gene for resisting it, or even several—the genes travel in packs. As we have used more and more antibiotics, the bacteria that survived were the ones that had these genes and could defend themselves, so they have become more and more common—there are few clearer illustrations of how evolution works. Antibiotics are prescribed at doses and over times that should kill all the bacteria, but even then, resistance can emerge. Infections resisted the "wonder drug" penicillin just three years after it was first widely used in people.

But misused antibiotics, as when patients demand antibiotics for ordinary flu, promote resistance even faster. So do the low doses of antibiotics given to cattle, pigs, and poultry to make the animals grow faster. There have been choruses of denial from livestock industries, but the science is clear: this practice contributes to antibiotic resistance in bacteria that cause human infections. Researchers have tracked it, as they say, from farm to fork. The European Union has banned antibiotic growth promoters, which demonstrates that modern animal production doesn't need them, but the US has been slow to give them up, and the drugs are used massively as livestock production booms in South America, Asia, and Africa.

We really don't want to lose antibiotics, especially if we expect more pandemics. Few people realize what a huge difference they have made to human welfare. Indeed, few people reading this have not at some point—probably several points—had their lives saved by them. Have you ever had an operation, even something ordinary like having your knee fixed or appendix removed? You needed antibiotics to stop the bacteria that invaded your opened-up body. Have you or a loved one ever had cancer treatment? Cancer drugs suppress immunity, so you need antibiotics, or bacteria can kill you.

Have you ever had an abscess, an injury, dental surgery, a bout of bacterial pneumonia, a sexually transmitted disease like gonorrhea, or a common urinary tract infection? Antibiotic resistance affects all of them, and there are now cases of those last two that resist all known antibiotics—they're incurable. Mothers and babies used to die in droves from bacterial infections—where they can't get modern medical care, they still do. That used to be normal for all humans: a tiny cut could result in gangrene or sepsis. Now, if you get an antibiotic-resistant infection in a cut, this can happen again.

So invent better antibiotics, you say. Exactly. But for the same reasons we don't have better flu vaccines or treatments for coronaviruses, we don't have many new antibiotics hitting the market. In an investigation I did for *New Scientist* in 2019, I found that unlike a few years previously, researchers and research funding agencies had taken up the fight, and a lot of new kinds of antibacterial drugs are in development, including clever new approaches like harnessing viruses that infect bacteria.

But industry experts warned me that however good these were, they were unlikely to get the $1 billion worth of trials in people a drug needs to be marketed safely. Antibiotics, like flu vaccines, are not terribly profitable. People only take them for a week, whereas they go on buying blood pressure pills, arthritis drugs, or Viagra for years.

Moreover, new antibiotics should not be widely or aggressively sold, but ideally saved for infections that resist existing drugs to avoid encouraging resistance to the new drug. Yet it is when a drug has just come on the market that companies most need to start selling it heavily to recoup their R&D investment. And even when new antibiotics are the best treatment and should be used, doctors tend to try older, cheaper ones first. As mentioned earlier, there are proposals for methods of repaying companies through mechanisms other than sales, but none has been widely applied yet.

Thus, many Big Pharma companies have abandoned R&D for

antibiotics: 18 developed them in 1980; now only six have any kind of program, and they may not last. Fortunately, small start-up companies are developing new antibiotics, but in 2019, several went bankrupt. One in California, Achaogen, had spent the requisite $1 billion getting a drug that cures antibiotic-resistant urinary tract infections, Zemdri, onto the market. However, the company needed more money to market the drug and do more studies. Investors saw little chance of profit and declined. The drug was bought by another firm, but the company is gone, the researchers are off doing other things, and fewer inventors with a great idea for a new antibiotic are now likely to pursue that.

So we risk losing antibiotics. In 2014, a blue-ribbon commission in the UK reported that 700,000 people a year were already dying worldwide of antibiotic-resistant infections, 50,000 just in Europe and the US—but by 2050, this number could jump to 10 million a year, more than die of cancer and more than seven times the number who die in road accidents. That many deaths would slash trillions from global GDP. A pandemic virus would be proud of those numbers, if proud was something a virus could be.

This really matters to our risk of viral pandemics. We need new antibiotics to treat the bacterial complications of viral disease, especially in a pandemic where rocketing antibiotic use might lead to rocketing resistance to our existing drugs. The 2014 report forecast economic damage resulting in severe domino effects as we increasingly lose antibiotics: less money overall means less for new treatments for emerging disease, for stockpiles in case of a pandemic, for disease surveillance.

The good news is that the solutions we need to encourage new antibacterial treatments are largely the same as those we need to encourage pandemic precautions, such as regularly updated stockpiles of masks and ventilators, or drugs, vaccines, and tests for potentially pandemic viruses. You guessed it: what they all have in common is that, for all its virtues, the free market on its own cannot make them happen.

That means public investment. An organization called CARB-X has $500 million to invest between 2016 and 2021 to accelerate development of new antibiotics, similar to what CEPI does for vaccines. "One lesson from COVID-19: we need to invest today for tomorrow's pandemic," says Kevin Outterson of CARB-X. "What would a COVID-19 treatment or vaccine have been worth in 2018? Zero—the market would have seen no value at all. What would they be worth today? Off the charts. That is true for pandemic viruses and also for drug-resistant bacterial infections."

But like CEPI, CARB-X only gets products as far as initial trials. The problem is the big expensive push they then need to get onto the market. With Covid-19 drugs and vaccines, the need is obvious, so the money will be found. With antibiotics, the need is clear too, but even so, few new ones are in large-scale clinical trials. We may need to get creative.

One model is the US Project Bioshield, set up in 2004 after the anthrax attacks to help develop vaccines and treatments for germs that might be made into bioweapons. This was established not to provide initial R&D, but to get companies with a promising product through the "valley of death"—the long wait during safety and effectiveness trials before they can sell.

Another US agency, the Biomedical Advanced Research and

Development Authority, or BARDA, took it over in 2006, expanded the remit to pandemics, and has invested $1 billion in antibiotics. Unfortunately, that included $124 million for Achaogen, which didn't keep the firm from going under. But that just shows we need more consistent follow-through. Adalja observes that Bioshield cost in the low billions of dollars, far less than Covid-19 will cost. Preparation is always cheaper than reparation.

One can imagine some mechanism like that set up internationally for pandemic threats, with enough funding to get drugs—pandemic and antibiotic—and vaccines through trials. Guarantees from governments to buy the results would help, and working globally—which, after all, is what pandemic diseases and resistant bacteria do—would bring economies of scale. It seems at least worth a try. If we don't include antibiotics in any grand plans for pandemic preparedness, we are setting ourselves up for failure.

There will be lots of strife long before the dust settles on Covid-19 over what worked and what didn't and how this or that solution was too late or ignored. We should turn that to constructive use, if we can, to hammer out plans that actually will work next time. We have already learned a few things.

We need better rules for using the high-tech methods and smartphone apps now being devised to trace contacts and enforce lockdowns. Contact tracing especially is difficult and expensive, but vital as we have learned, and made vastly more effective using apps like the ones now being developed. There must be ways to use these technologies without demolishing people's privacy or enabling authoritarian streaks (or worse) in some governments. Watchdogs with serious teeth will be needed.

The lack of medical equipment has rightfully received much attention. It is to be hoped that countries will stockpile the things we have desperately needed during this pandemic: ventilators, protective equipment for medical staff, and masks for the public. Experience is a good teacher: the Canadian province of Ontario, where Toronto was hit hard by SARS in 2003, had extra ventilators stockpiled that it used for Covid-19. I wonder how often Ontario had to defend that stash against critics calling it a needless expense now that SARS was gone. In 2006, California governor Arnold Schwarzenegger made similar stocks of pandemic supplies; they were lost to a budget crunch under another governor in 2011, and the state is now short of ventilators.

Adequate testing has also been a major, unexpected issue. Another idea would be pre-pandemic contracts with test manufacturers to be activated in an emergency. South Korea signed agreements with test makers on the fly early in the Covid-19 pandemic, which let the companies develop and start mass-producing tests in days and also to use them and put them through validation trials at the same time. That famously enabled South Korea to contain the virus fast. Other countries could follow their lead and pre-arrange such agreements with test companies and other suppliers of emergency pandemic goods.

Meanwhile, individuals should not have to choose between spreading a pandemic and feeding their families. Even before this pandemic, a research found that paid sick leave, so employees do not engage in "presenteeism"—going to work sick—ultimately saves companies money. In the Covid-19 pandemic, it saved lives. Guaranteeing that right for working people even in the gig economy is possible, says the UN's International Labour Organization, and would increase resilience to infectious disease in places where it is not already an unquestioned labor standard.

Of course, all these ideas—pandemic stockpiles, global surveillance, flu vaccine, sick leave—cost money. But put it in context. The GPMB assembled some sobering figures. Zika cost the Americas some $20 billion, including care for the many handicapped children it left. The 2014 Ebola epidemic in Africa cost the world $53 billion. A lot of the financial loss to the three countries directly affected by Ebola was due to medical services not performed— vaccination, childbirth, malaria treatment—because of the disruption of medical services. That is already happening with the Covid-19 pandemic, as crucial polio vaccination is suspended in Pakistan and cancer deaths in England are expected to rise by 20 percent this year due to disruption in medical service and people avoiding hospitals. Epidemiologists at Imperial College in London calculated in May that because medical services are pre-occupied with Covid-19, deaths from HIV, TB and malaria over the next five years in low and middle-income countries would rise 10, 20 and 36 percent respectively.

SARS cost the world $40 billion. The 2009 flu pandemic cost $55 billion. A repeat of the 1918 flu pandemic would, it's been estimated, cost the modern economy $3 trillion, nearly 5 percent of global GDP, triggering a worldwide recession. Now some economists predict that the lockdowns needed to prevent deaths from

Covid-19 might lead to a full-blown depression. At least as a result of Covid-19, no one can dispute that the cost of pandemic prevention and preparedness is just a drop in the bucket compared to the cost of an actual pandemic.

Seth Berkley at GAVI is one of those people you meet in Geneva who is very clear-eyed about the world after years of working internationally. Three years ago, when I was writing about preparing for pandemics for *New Scientist*, I asked him how to make countries take the problem seriously.

Seth is American, and he observed that the US maintains three kinds of nuclear weapons—airborne, land-based, and submarine-launched missiles—so it will have one deterrent left if the other two fail. "The chances of that happening are very small, but we spend tens of billions of dollars a year to keep that running," he said. His point was that, if it's worth investing $49 billion a year to ensure you can respond to an improbably successful nuclear attack, surely we could invest in improving our response to an increasingly probable pandemic. Yet this year, the whole world will spend only $2.4 billion, 5 percent of the annual cost of the US nuclear deterrent, on the WHO.

He also said something I don't think I really understood until Covid-19. "The really big problem is appreciating what is at stake," he told me. "If people understood the risk, they would want to be sure systems are in place to deal with it. The costs of doing that are trivial compared to the cost of ignoring it."

We all know that now. The only question is whether we will forget about it again, after Covid-19, or whether we will finally be able, together, to act on it.

## CHAPTER 7

# Things Fall Apart

> "So, it turns out the most important jobs
> are not the bankers, the brokers, or the
> hedge fund managers. It's the doctors,
> the nurses, the hospital porters, the A&E
> administrators, the bin men, the teach-
> ers, the carers, the supermarket shelf
> stackers…Who knew?"
>
> —T-shirt from 2020 about Covid-19

A few years back, I started looking at what would happen if another plague like the Great Pestilence of 1347 hit the world. Dubbed "the Black Death" by later scholars, the medieval plague killed a third or more of all people in Europe and spread into Asia. I was writing about potentially pandemic pathogens with high death rates and wondering, what if?

Now let me say right at the start that I am not even remotely suggesting that Covid-19 is showing any sign whatever of becoming that bad. If anything, the more we learn, the milder it often seems to be—except, of course, in some people, when it isn't. But there are viruses out there that are a lot worse than Covid-19, and as we will discuss, it is not at all a foregone conclusion that they will start playing nice if they go pandemic.

In any case, what I discovered by asking what might happen if another Black Death hit us turns out to be very relevant to what we are already going through with Covid-19, even though it has been much less lethal. The link, as with many apparently intractable problems, is complexity.

European civilization did not collapse from the mass death of 1347, even though the disease circulated, with smaller outbreaks, for the next 300 to 400 years. Some historians even think the resulting labor shortages shook up the rigid feudal system and triggered changes that led to the modern era. The key was the simplicity of the agrarian society it struck. In such systems, nine in ten people are subsistence farmers, producing just enough to feed themselves, plus a meagre surplus that, in medieval Europe, fed a few aristocrats, churchmen, and towns. Most deaths took out a peasant, and therefore a producer, but also a consumer, so had little net impact on society. Non-food producers who depended on peasants were taken out at about the same rate. Even kings were replaceable.

But, as Joseph Tainter, author of *The Collapse of Complex Societies*, told me, in 170 CE the Roman Empire was hit by an almost equally deadly plague that he believes tipped Roman civilization into a death spiral. The difference was the empire's large urban populations, not equaled until modern times, and the networks of grain shipments and taxes and armies that supported them. Losing a third of the population meant grain production and taxes fell, the army suffered, invaders Rome would once have easily repulsed made inroads, grain and taxes fell further, and so on. Ultimately cities—the *civis* that was Roman "civilization"—largely disappeared. Decline led to fall.

The difference was complexity. Defined very simply, a complex system is one in which many elements interact closely and feed back on each other—change one bit, it changes another, which changes a third in a way that reverses the first change a bit but also affects a fourth, depending on what a fifth is doing. The important thing to know about complex systems is that they behave very differently from the linear, mechanical systems we are more familiar with, where if you put something in one end, you get a predictable response out the other.

In a complex system, if you change one bit, you might get a completely disproportionate response you were not predicting, because you don't know the states of all the components at that precise moment or how they all affect each other. The famous butterfly effect, where the flap of a butterfly's wings in Brazil could set off a tornado in Texas, reflects early efforts to model the weather, a complex system where tiny differences in starting conditions can create huge differences in outcome. These are called nonlinear effects. This happens in all complex systems. A large change can also have small effects—up to a point.

This matters, because complex systems have a few more universal properties. Complexity can only be maintained with energy. The laws of thermodynamics, the most fundamental laws of nature, make it quite clear that, in strict scientific terms, there is no free lunch. To maintain a system more complex than random atoms—

you, for example—you need to spend energy on it. For you, this comes in the form of food. You process the energy and materials bound up in the orderly structure of your sandwich to build and maintain the orderly structure of you, except for some energy lost to friction in the system. And something else had to do that to produce the materials in your sandwich. No free lunch.

Moreover, complex systems tend to remain stable in the face of the normal range of conditions they evolved to handle, a property called resilience: disturb the system, and complex, adaptive readjustments will keep it on an even keel. This isn't magic. Complex systems evolve over time, by trial and error, and tend to self-assemble rather than being designed from outside: the enormously complex network that manages to deliver an amazing array of food to a big city every day is an example. Resilience evolves in such systems for the same reason anything evolves: because it can, and it works.

But if you push such a system outside the conditions it evolved with, resilience can vanish. A small change can flip it into an alternative stable state—the famous tipping point. A straw breaks an overloaded camel's back. A few tiny bacteria that produce deadly botulinum toxin, hiding in that sandwich you just ate, can kill you. So can a virus.

Society is a system that gets steadily more complex, says Tainter, because whatever we do, we encounter problems we have to solve. We start growing crops to have more reliable food, but the rains sometimes fail, so we dig irrigation canals. Those silt up, so we invent dredging. They silt up more, so we have permanent dredging crews, and they don't farm so we give them food other people grew. Disputes arise, so we invent ways to record who gave and got what, then a class of people to keep order, who must also be fed. You can see where this is headed.

Human history is a long saga of people learning to harness ever-increasing amounts of energy to maintain ever more complex, ordered systems, punctuated by periodic collapses—the Romans, the Maya—when civilizations became more complex than they could maintain, with the energy and technologies they had, in the face of changing conditions. At that point, small stresses sent overstretched social systems into a rapid downward spiral, which ended with major losses of people and social organization, as one stable complex system made a rapid nonlinear descent to a less complex one. But after a setback, humanity always innovated and rebuilt, a little bigger and more complex than before.

This process is integral to how we should understand pandemics. We now live in the most complex civilization the world has ever seen and the first to encompass the entire planet. Many believe that this makes us resilient to shocks. But, say the complexity theorists, the more complex systems get—the more tightly coupled their component parts, the faster and denser the communication and transport links that keep them all coordinated, the more closely each part relies on many other parts—then the more rigid the system gets overall, the less resilient, the more likely to collapse.

Moreover, complex systems—natural ecosystems as much as human societies—tend to become more efficient, with more specialized components and fewer redundant linkages, because that saves money or energy. Thomas Homer-Dixon, a Canadian expert in complex systems and author of *The Upside of Down*, notes that a mature forest may have one kind of bacteria fixing its soil nitrogen, whereas at an earlier stage of development, it had a dozen.

Similarly, protective medical gear and the active ingredients for common, emergency drugs used to be produced widely. Michael Osterholm is an epidemiologist who has studied the possible impacts of pandemics. He told me that now, a few factories in China make nearly all of these vital supplies, as the global industry takes advantage of low labor costs and economies of scale. This is efficient. Hospitals rely on constant, just-in-time deliveries of these items, too: keeping stocks costs money, so this is also efficient. During the early days of the Covid-19 pandemic when much of China was affected, there were fears deliveries would stop, either because China needed more of these things than usual or because factories or shipping might shut down as employees were quarantined. If things had gotten much worse or the shutdown had lasted longer, they might well have. There were no alternate sources. Efficient coupling between parts of the system would have led to a breakdown.

Homer-Dixon says increasing complexity makes societies more resilient only up to a point. Connections between villages might mean one comes to the other's aid in an attack. But as the villages become more tightly coupled, both may suffer when one is attacked. A loose network absorbs shock; a tightly coupled one transmits it.

That is happening in the Covid-19 pandemic. Countries go into lockdown; people stop shopping, traveling, and producing; and the effects ricochet through a tightly coupled global economy. The global supply chains of money, materials, people, energy, and component parts that underpin industries falter and break. Airlines go under as they are not set up to weather even a temporary disappearance of travelers. Malaria worsens in Africa as insecticide and antimalarial bed net deliveries falter. Microcredit that underpins small businesses throughout the developing world defaults because payment collectors are locked down, causing ramifications throughout an economy.

The number of people facing starvation threatened to double in April 2020, warned the World Food Programme, even though the same amount of food was available. Lockdowns meant poor people, from tuk-tuk drivers to cleaners to food vendors, could no longer earn money to buy it—and this happened just as restricted global transport made it hard to get donated food to them.

Just the fact that an outbreak in China went pandemic as quickly as it did is a testament to the tight coupling in our global system. For human viruses, the vector is people and airplanes. Scientists tracked this pandemic using computer models and databases of global air passengers. Alessandro Vespignani of Northeastern University calculated that the countries at highest risk of importing a case of Covid-19 were in Asia, followed by North America and Europe; that is exactly how the virus traveled. Vittoria Colizza of Sorbonne University in Paris calculated that the African country most likely to import a case was Egypt, followed by Algeria. Those countries, in that order, got Africa's first cases.

The fact that the world is a complex system helps explain how this pandemic happened. First, it means our system has a management problem. People tend to see things in a simple linear way. That's not a criticism—we can't usually control anything but a few, simple, direct interactions within our complex social system. So faced with a problem, those are the solutions on offer. We cannot always anticipate how the rest of the complex system will impinge.

Chinese medical authorities said, we had a close call with SARS and we have bird flu, let's have doctors signal any unexplained pneumonia, and we'll see any clusters of cases faster. Scientists said, we have a problem with animal viruses infecting people, let's swab a lot of animals and see what they're carrying. Pandemic planners said, if we have a flu pandemic, we'll need ventilators and masks, so we'll stockpile them. All great ideas, and it was a good thing people did or tried to do all of them.

But complex interactions took an unexpected hand. In December 2019, when it was clear the unexplained pneumonia wasn't bird flu, for some reason Wuhan doctors were told not to use the alert system. Scientists found a virus much like Covid-19 in bats and warned of its pandemic potential, but that didn't nudge research agencies to fund renewed work on coronavirus vaccines. The 2008 market crash—itself a textbook product of complexity and tight coupling in the global financial system—triggered government cuts that tightened health budgets. Then the 2009 flu pandemic was less than catastrophic. Result: hospitals, with few exceptions, did not get the pandemic stockpiles envisioned in 2006.

Western countries have been talking about pandemic preparedness since bird flu rang alarms in the early 2000s. This was especially true in the US, which was widely expected to be the country best prepared for something like this. But when Covid-19 hit the US, the plan was largely abandoned, while unexpected complications set in everywhere. Health workers didn't have enough protective gear and ended up sick or in quarantine. Insurance rules meant people initially couldn't afford to get tested. For weeks, they couldn't get tests in any case because of problems with one test at the US CDC in Atlanta. Employees with no paid vacation came in to work, hoping it was just flu. The virus spread earlier and farther than surveillance systems could detect, partly due to years of cuts to public health.

Pandemic planners actually warned about many of those problems. But no one could change enough of the system to head them off, and when a severe pandemic didn't materialize after years of warnings, leaders lost interest. We can't prepare a complex system for events like pandemics with small, linear solutions to local parts of the problem; we also can't prepare when we lose focus on any given risk after a few years.

So is there any hope? Actually, there is. We will look at potential solutions in the next chapter. For the moment though, let's see how high the stakes really are. If we know how bad things can get, we'll know just how much we need to get cracking on those solutions—preferably before something worse goes pandemic.

For starters, how bad can a pandemic pathogen get?

That isn't as simple a question as it may appear. There is a widespread belief that when a new disease learns to spread among humans, it becomes less lethal. Many people believe there's an automatic trade-off between lethality and transmission.

This much we know: to survive, the pathogen needs to get into another host either before you die and take it with you, or before your immune system kills it. So it is often good for the pathogen if it doesn't kill you all at once, as it needs you to spread its progeny, by coughing and sneezing if it's a respiratory virus. As a result, milder pathogens *may* win out over nastier ones as they adapt to humans. But this has been generalized into a broader conclusion: pathogens that are new to us are often severe at first because they haven't figured us out yet, but if they persist and spread widely, they *always* evolve to become milder so we will live long enough to spread them.

This is widely accepted as dogma. In 2005, near the peak of interest in pandemic preparedness, I was at a meeting on the topic at a very upscale British think tank and found myself in conversation about H5N1 bird flu with a then-member of COBRA, the top advisory committee called in by the British government in emergencies. I asked how we prepare for an H5N1 pandemic, if one emerged, given that it kills 60 percent of the people it infects. He looked me dead in the eye and said, "Don't worry. If it goes pandemic, that death rate will fall. These diseases always become milder." What, just like that?

The 2018 UK flu pandemic plan shows how entrenched this assumption has become. It is based, as a "reasonable worst case," on flu that kills 2.5 percent of cases, like the 1918 Spanish flu. It acknowledges that H5N1 bird flu kills 60 percent, but simply states that if H5N1 becomes able to transmit between people, it will kill far fewer. While deaths above 2.5 percent "cannot be ruled out," it says, a pandemic H5N1 "would be expected" to kill around 2.5 percent. Moreover, it tells local authorities to plan for a 1 percent death rate, as, it states, a more lethal virus that attacks many people is simply unlikely.

Recent history doesn't do much to support the idea that diseases from animals always become mild as they adapt to spreading in humans. Consider HIV. It spreads before you get sick, so it's no problem for the virus that after a few years, without the right drugs, 100 percent of infected people die. As we have seen, it hit humanity in the early 20th century, it eventually went very much pandemic, and it's shown no sign of getting nicer in all that time.

Or look at bird flu. It is a benign gut virus in ducks, as it really does need the duck to swim around pooping it out for a while if it's going to reach another duck. But once it gets into a henhouse, it often mutates into a lethal, highly contagious infection. There is no shortage of chickens, and a simple mutation, turning it into a virus dubbed "highly pathogenic," allows it to replicate explosively and get into the next bird before other viruses. That this kills the chicken makes no difference to the virus. In fact, this mutation is beneficial enough for bird flu, at least in the short term, that this often happens when duck flu viruses of all kinds, not just H5N1, end up in a henhouse. In this case, getting a lot more deadly, at least in chickens, works out just fine for the virus.

Most worrying is that, as we saw in Chapter 5, a few mutations made H5N1 contagious between mammals, but it seemed to remain just as deadly. There didn't seem to be any trade-off between transmission and virulence (severity). This is all the more worrying because another bird flu virus, H7N9, with a 30 percent death rate, had three of the five mutations required, occasionally spread a bit between people, and viruses with the "highly pathogenic" mutation could spread between ferrets and kill them, just by being inhaled. No apparent trade-off there.

Now, let me repeat that this does not apply, at the moment,

to Covid-19. As virologist Ian Jones at the University of Reading observed in April 2020 on Britain's Science Media Centre, which sends commentary from scientists to journalists, the virus is under no particular pressure to evolve. "At our cost, the virus is doing well enough colonising the human population. I don't see the drive for it to get nastier anytime soon."

On the other hand, random mutations constantly occur in these viruses, and if one were to enable the virus to spread even better in us, it might well become more common in the virus population, or be "selected for" in evolution-speak, as spreading is what benefits viruses most. As I write, mutations are accumulating in the virus that causes Covid-19—not surprising after four million infections in humans—but we don't yet know if any are actually changing the behavior of the virus.

If that happens, and that mutation also affects virulence, then the virus that emerges might be either more or less virulent, whatever works best for the virus. Let's bust this myth: it is not a hard and fast rule of pathogen evolution that it must be less, or that there is always a trade-off between virulence and transmission. It is a bit worrying that people who are not primarily pathogen evolution experts, but are involved in pandemic response, seem to believe it is. It is also worth remembering that the Covid-19 virus's waltz with humanity is just beginning.

Meanwhile, it's not just a question of how viruses act on their own: some vaccines can increase virulence. Andrew Read of Pennsylvania State University has done research with several diseases, including Marek's disease, a common plague of chicken farms, to see how vaccination affects the virus's evolution. He found that if a vaccine keeps the virus's host from getting sick, but still allows the virus to persist and spread—like the poultry vaccine for H5N1 did in China—it can select for a more virulent virus.

This is because it is actually true that pathogens can be too deadly to spread: they get only so virulent, then if they get nastier, they kill off their hosts too fast, and die out. The occasional virus like that might emerge, but it doesn't get far.

"Leaky" vaccines, though, induce immune reactions in the host that keep the virus at low levels, so an infected host doesn't die—but the virus doesn't completely die out either. If that happens, those viruses can become extra virulent, because they don't have the problem of killing off their host if they do. And changes in the virus that increase virulence—faster replication or faster invasion of the host's cells, for example—might be helpful for a virus in a vaccinated host, allowing it to persist and spread a bit better despite the host's immune reactions. Now if that virus reaches hosts who aren't vaccinated, it will be more lethal than usual. That is exactly what happens with chickens vaccinated for Marek's.

This could be a concern if Covid-19 vaccines are "leaky." "There certainly are plausible scenarios under which leaky vaccines could drive the [Covid-19] virus to increased virulence," Read told me. "I can also see scenarios where it could go in other directions." It all depends on what works for the virus. If we develop a "leaky" vaccine for Covid-19, and some strains became more virulent, it could cause trouble. We are unlikely to vaccinate everyone—we never have with any other vaccine, even when we eradicated smallpox. If a virulent mutant of Covid-19 circulated silently among vaccinated people, then reached people who were not vaccinated, it could be bad.

We will have to keep this in mind as we develop Covid-19 vaccines. The virus is likely to keep circulating, at some level, everywhere until we have a vaccine and use it widely. That means countries will need to keep testing and interrupting chains of transmission to contain it—or if they fail to, they will have more large outbreaks requiring social distancing. Either way, it will be expensive, so there will be enormous pressure to widely use whatever vaccine we develop that seems to work.

Now, imagine we find such a vaccine, but then someone says: you can't use that, it might cause the evolution of a more severe strain of Covid-19. That might be a hard sell.

Meanwhile, adapting to humans seems to have resulted in increased, not decreased, virulence in Ebola. The 2014 West African Ebola epidemic was by far the biggest ever, with nearly 29,000 known cases and a recorded 11,310 deaths, although careful analysis suggests those numbers are incomplete and the real death rate was closer to 70 percent, which is more typical for Ebola. That's compared to 315 cases, though, in the Congolese city of Kikwit in 1995, even though like the 2014 epidemic, it was also in an urbanized area. In the past, Ebola has actually been fairly difficult to catch, and outbreaks have been limited.

True to form, in 2014, the virus initially moved slowly from Guinea to Sierra Leone. Then something changed: it acquired a mutation in the part of the virus that latches on to human cells. After that, the virus spread much faster, re-invaded Guinea, raced across Sierra Leone and Liberia, and almost got loose in Nigeria, which stopped it by using a disease surveillance system built for the polio eradication program. After it acquired the mutation, all the subsequent viruses in the epidemic had it. Jonathan Ball, a virologist at the University of Nottingham in the UK who tested the virus, says it was almost certainly an adaptation that allowed it to spread better in people—and it was not milder.

Jeremy Luban of the University of Massachusetts, who did similar tests, agrees that the mutation showed all the signs of being an adaptation to people: it emerged after unprecedented circulation in people, allowed the virus to bind better to human cells, and then dominated the rest of the epidemic. Andrew Read observes that Ebola spreads when severe cases—and recent corpses—shed the virus in body fluids, so a more severe virus will get more chances to spread, making severity an advantage for the virus and, therefore, increased severity likely to evolve.

The mutant seems to have disappeared with the stamping out of the 2014 epidemic, but it could emerge again if another outbreak has similar numbers of human cases. The bottom line is that Ebola was apparently adapting to us, and getting nastier, not nicer.

How did the belief that pathogens always become nicer when they start spreading more easily become so widespread? It started, says Read, with myxomatosis in rabbits. The virus causes a mild disease in animals in the rabbit family in its native South America, but it causes a lethal illness in European rabbits. It was released in 1950 in the Murray Valley of Australia to control European rabbits, which had become an invasive pest. The same thing was done by landowners to cut rabbit numbers in France in 1952 and the UK in 1953.

The story you hear is that the virus rapidly evolved to become a mild disease, as the original strain killed too many rabbits too fast, ran out of nearby hosts, and failed to spread, while milder viruses thrived. This is the story people cite when they say a virus that learns to spread well in us will become milder.

In 2015, Read and his colleagues did a review of the research into what actually happened. Here's what they found. A few months after its initial release, the virus exploded across southeastern Australia. The toll was incredible: it killed some 95 percent of rabbits in farming areas. Much the same thing happened in Europe: the British children's classic *Watership Down*, about a band of rabbits, called it "the white blindness."

A year or two later, Australian virologists started finding slightly less lethal viruses among surviving rabbits. These were not, by any standard, mild. The most common killed 70 to 90 percent of laboratory rabbits, which are the same species as the wild ones. But they took longer to do it than the original strain, so they did give the virus more chance to spread. The researchers also found strains that killed fewer than half the rabbits, but they didn't transmit well, as the rabbit's immune system tended to kill them first, so these were rare. Much the same thing happened in Europe.

The die-off put a massive "selection pressure" on the wild rabbits. Basically, the few remaining rabbits survived because they had genes that made them resistant to the virus. While the virus persisted, having those genes was a real advantage, and rabbits with them rapidly became the majority. Being rabbits, they replenished the population, and the new ones were resistant to myxomatosis: seven years later, the virus was killing only a quarter of the wild rabbits. But it was as lethal as ever in lab rabbits, which hadn't been selected for resistance.

So the virus did become somewhat slower to kill, if not exactly mild—and the disease overall became less lethal because the few surviving rabbits gave rise to a population that resisted it.

Then in 2017, Read and his colleagues discovered the virus was fighting back: now it gets around the rabbits' resistance by attacking the immune system directly and spreading more readily. So again, adapting to its host meant becoming more lethal.

Scientists don't talk about myxomatosis becoming mild so it can spread, as though the virus somehow surveyed the situation and decided it had best do that. They talk about an arms race between the virus and the host. Myxomatosis did become a bit less lethal, but that was only after it had killed nearly all the rabbits. There was no reason for it to become any milder until that happened, and even then it was still pretty lethal. The disease overall looked milder, because the surviving rabbits were resistant. Then the virus got worse again.

Somehow, I don't think that was what that COBRA guy had in mind. I hope the people doing our pandemic planning are thinking carefully, and knowledgeably, about what really happens when viruses adapt to us. And I hope Covid-19 vaccines aren't leaky.

If we are hit by a much more virulent virus, how bad could things get? We have established that our globalized, interdependent world is surprisingly fragile. The domino effects of a pandemic on global production and trade can severely damage an urban economy dependent on just-in-time goods and services from the rest of the planet. That much we are learning from Covid-19.

But what is the impact of losing people—not temporarily to lockdowns, but permanently? We are losing people now, of course, but on nothing like the scale we could from a really bad pathogen—and we have seen that a virus with a higher death rate than Covid-19 can go pandemic. Beyond the immediate tragedy and sorrow, what is the impact on a complex, fragile world like ours of a lot of us dying?

It isn't necessarily self-evident. A lot of our problems, as we have seen, stem from the way we are managing our unprecedentedly huge numbers, including the pressures of poverty and economic competition that lead to our encountering new pathogens. A friend of mine was once listening to me going on about bird flu and pandemic threats, and said, "Look I don't want to sound callous or anything, but, well, wouldn't it be better in some ways if there were fewer of us?" That was the question I set out to answer.

Our globalized, industrial society is effectively arranged vertically, with almost everyone totally dependent on support by many other human subsystems called critical infrastructure: housing and heating, food production and distribution, water supply and sewage management, public health, transport systems, security services, telecoms, banking, shops supplying essential goods and services, electric power. To some extent, all the subsystems rely on each other. With all these complex interdependencies, this makes us, basically, a big game of Jenga: pull a few pieces out anywhere in the system, and the rest can fall.

What may not be immediately apparent is that a lot of the most important pieces are the people themselves, says Yaneer Bar-Yam, head of the New England Complex Systems Institute. It isn't obvious, he told me, but research with complex systems shows that the more complexity rises, the more individuals matter. If a more lethal pandemic took out more of the key people running our critical systems, the impact could be pervasive.

Some industries are hubs—like bats in an ecosystem, a lot of other parts depend on them. Industries in turn depend on their workers. In 2000, a strike by truck drivers blocked nearly all gasoline deliveries from Britain's oil refineries for ten days. Public transport collapsed, grocery stores emptied, hospitals ran minimal services, hazardous waste piled up, bodies went unburied. The government had to step in. A subsequent study predicted economic collapse in Britain if all road haulage, not just fuel deliveries, was shut down for only a week.

Today, we all depend even more on just-in-time deliveries: if the trucks stop because drivers are locked down, or sick, or dead, or caring for sick family, cities will rapidly have no food, vehicles won't have fuel, food in depots will rot. In the future, if deliveries depend more on automated systems, trucking may not remain as vulnerable—but the principle remains that if certain hub industries are paralyzed by loss of people, the impact can be far-reaching. There will be other choke points that depend on people: doctors and nurses, engineers who run power grids or essential manufacturing, or global supply chain managers are not all readily replaced.

Even transient absences of key workers can cause snowballing problems. During Covid-19 lockdowns, oil refineries are shutting due to plummeting demand as air and road traffic fall. In a pandemic with a high loss of people, absence of workers at oil refineries starts becoming a problem. The current UK pandemic guidance for the natural gas industry predicts that anything more than staff absences up to 30 percent for a month "would be problematic," whereas an absence rate of 45 percent—or possibly less during peak demand in winter—could trigger a Gas Deficit Emergency, with some users, like factories and homes, shut down.

As always in a complex system, the problem doesn't stop there. Oil refineries also produce transport fuel, and lack of that stops deliveries, including of coal, on which much electricity in some countries still depends. That's where things really start to collapse, Michael Osterholm told me. Failure of electricity will cripple subsystems from lighting to ATMs to refrigeration to pumping drinking water, and electric power is needed for mining coal or pumping oil for generating the electricity itself. It gets worse from there.

Truck drivers and refineries are only two sectors where this domino effect could start. Once one part of a network of interdependencies wobbles, the rest is at risk.

The Covid-19 pandemic does not have a high rate of sickness and death, so large percentages of the population are not being disabled by disease. But the social distancing measures being used to slow its spread are stopping much economic activity. Tellingly, workers deemed essential to critical infrastructure are exempted from this everywhere. There are some cogs in our system that we really need to keep turning.

The official US list of essential workers for this pandemic makes fascinating reading. A random selection: health workers at all

levels including cleaning staff; building security staff; food workers; crop pickers; miners; armored cash transporters and ATM
servicers; powerline repair people; truck stop operators; grocery
store workers; the people who cut tree branches away from over-
head electrical lines; sewage processing plant workers; road repair
crews; bus drivers; plumbers; waste disposers; telecommunica-
tions repair people; IT workers who maintain the internet; metal-
workers; chemical workers; laundromat staff; janitors…

Of course, there are also judges and lawyers, doctors and power
plant engineers, cyber-defense experts, some clergy, and other
white-collar workers on the list, but a glance shows that a lot of
critical infrastructure depends on low-income people. It has long
been known that low-income people are more likely to die of infectious disease generally, due to underlying poor health and, in some
countries, lack of access to medical care. The worse outcomes with
Covid-19 among disadvantaged racial minorities in the US suggest that, unsurprisingly, this applies to pandemic disease too. A
massive British study released in May found the poorest patients
with Covid-19 were almost twice as likely to die as the richest, and
it mostly wasn't because they had pre-existing illnesses. Meanwhile, say economists, income inequality already showed every
sign of continuing to deepen, while research has found that epidemics disproportionately hurt the poor, making things worse.

That means more people in poverty, which in turn means more
people who are extra vulnerable in a pandemic—including many of
the people responsible for critical infrastructure. Some—for example in the meat-packing industry, where social distancing has been
absent and Covid-19 has struck hard, but also in health care and

many other sectors—are immigrants, some of whom are undocumented and less likely to have access to health care, and many of whom must work despite illness.

Greater vulnerability among lower-income people worsens the spread and impact of a pandemic in the most critical parts of the complex system: firefighters, paramedics, police, care workers, the people who produce everyone's food, drinking water, electric power, the list goes on. The less those people can withstand a pandemic, the more the system that supports everyone is at risk of collapse. More inequality, and more poverty, means more risk.

No pandemic plans that I have been able to find seem to take into account the domino effects propagated through our complex critical systems simply by the deaths of critical people within them. Most engage in wishful thinking about death rates. The UK flu pandemic plan with its assumption of less than half the death rate of 1918 is typical; Tim Sly, an epidemiologist at Ryerson University in Toronto, says he has never found one that assumes an even worse death rate than that of the 1918 flu, even though we know there are deadlier flu viruses. Perhaps planners assume that if viruses go pandemic, they will become mild; perhaps the alternative is simply too harrowing to consider.

Tabletop simulations of pandemics with real-life government officials and industry bosses find many aspects of society rapidly collapse as unexpected consequences pile up—and, I was told by people who conduct such simulations, participants always discover, to their surprise, that their key personnel actually are the critical infrastructure. If a highly lethal pandemic takes out a lot of people, the consequential failure of our support systems may go on to take out more.

We rarely think about just how precarious these systems are, but the evidence is everywhere. A common saying in security circles is that a city is never more than three or four meals away from anarchy: as food prices on the global market rose in the run-up to the financial crisis of 2008, there was rioting in many places. And systems support each other: for example, if chlorine for water purification is not delivered just as power outages make it hard for many to boil water, waterborne diseases could result. Often we don't see the problem until it's in our faces: New York discovered after Hurricane Sandy that high-rise residences dependent on elevators became traps for people less able to handle many flights of stairs when the power went out, while even hospitals with generators failed to cope.

Countries rely on foreign deliveries for everything from milk cartons to pharmaceuticals, and in a bad pandemic, shipping will falter even more than it has with Covid-19. Even in this pandemic, people on large oceangoing vessels, be they cruise ships or aircraft carriers, are turning out to be at increased risk for contagion. Container ships are nowhere near as heavily crewed or luxurious, but they still need people.

A pandemic with a high death rate could trigger a lot of this kind of domino effect. Another thing about complex systems is that they lose complexity rapidly, but gain it back with difficulty, if at all. Partly, this is thermodynamics: the first process releases energy, whereas the second requires it. But in addition, as we mentioned, the study of complex systems has revealed that these systems settle in stable states, from which it is hard to dislodge them.

And that brings us to collapse. Collapse is tipping from one state to another with less complexity, providing fewer services and able to support fewer people. If the collapse of various subsystems of our current society propagated globally, sweeping up other subsystems in the process, eventually some countries, industries, or economies could collapse as well and have trouble getting back up.

The more this encompasses people and life support systems, the more likely the collapse is to be existential, for some or all of us. Such a dramatic event might seem unimaginable, but Covid-19, even though it is, let us repeat, not nearly that drastic, has laid bare just how interconnected and fragile some of our systems are. UN Secretary-General António Guterres called the pandemic a wake-up call. "We have an opportunity now to do things differently," he told the BBC. "It is clear the world is too fragile in relation to the global challenges we face. That fragility was demonstrated obviously with the pandemic." It will be even worse, he says, with climate change.

On the bright side, an understanding that we all depend on complex systems can help us prepare for some of the challenges Guterres mentions, including the next pandemic. That means making our systems less fragile—but the right answers may not always be the most obvious. Over the past few decades, many jobs in the traditional industrial heartlands of Europe and North America have been lost to "offshoring," the movement of industries to fast-industrializing countries elsewhere. There is now talk of rolling that back and "reshoring" in some industries, especially those that are vital in a pandemic, to shorten fragile supply chains.

But that might not always be the best thing to do, warns Shannon O'Neill of the Council on Foreign Relations, a US think tank.

In many cases, such a forced move will lose those industries the advantages of scale and labor costs they gained by moving, resulting in rising costs of goods for ordinary people, by some—$10,000 a year on average for US consumers, a significant portion of average incomes. This would mean hardship for some. The disappearance of those industries from the largely-developing countries where they are now would mean hardship for many more.

Moreover, it could be prohibitively difficult to re-assemble a complex system like manufacturing in a new place: O'Neill cites an effort by Apple to make MacBook Pros entirely in the US in 2013, which failed because one type of screw could not be sourced locally. It almost sounds like the old saying: for want of a nail the horseshoe was lost, for want of the shoe the horse was lost, then the rider was lost, and so on until the kingdom is lost. Perhaps we have always had an instinctive understanding of complex systems.

Instead, O'Neill suggests putting more redundancy back into globalized industries, making them more resilient. It will cost, as efficiency was adopted in the first place to save money. But it may well be less expensive than dealing with collapsing supply systems in the next pandemic. Shipping analysts were already saying in April 2020 that they expect industries to now diversify their suppliers—even if that means higher costs. Of course, shipping would be the beneficiary—and carbon emissions could rise.

Homer-Dixon agrees that we need more redundancy in the system, but also less overall complexity, to increase the "slack" in our support networks that can absorb shocks. That could, he says, mean cutting international travel, simplifying global supply chains, and indeed bringing some crucial production closer to the final users, or at least putting it in more places.

He says it isn't just our connectivity that puts us at risk, but also our uniformity, not just biologically as humans but culturally, in our food, ideologies, social media, finances, consumerism, even our antibiotics. If we have the same responses to perturbations everywhere, we risk disaster everywhere if one goes wrong. "Diversity, often a key feature of complexity, can be highly beneficial," he told me. The problem is not complexity per se—it's whether it leaves you more, or less, vulnerable.

Too much uniformity sets us up for cascading failure—and also synchronized failure of apparently independent subsystems. Homer-Dixon led a group of leading complexity experts that reported in 2015 that the apparently separate crises of 2008–2009, when both food and oil prices rocketed just as a US mortgage crisis triggered financial turmoil, were deeply linked.

The global economy has never gone back to the way it was before the crisis and has apparently found an alternate stable state. Homer-Dixon predicts that this pandemic will similarly be "a global tipping event, in which multiple social systems flip simultaneously to a distinctly new state." And, he says, if we don't start managing the problems raised by our complexity, we will get more of them, with ever-higher destructive force. The potential problems of a bad pandemic pale beside some of the possible impacts of climate change.

A pandemic of a far worse disease than Covid-19 could be one of them. Deaths of large numbers of people would be hardship enough—but it would have the further insidious effect of taking out many of the key lynchpins in our complex global systems, entailing further losses. No, is the answer to the question my friend asked: we can't lose a lot of people and put less pressure on the planet, while things continue as before. We would lose a lot of people, then a lot of everyone else, and nothing would continue as before.

So what would be the result of facing something like the Black Death in our modern society? We are not as resilient as Europe in 1347. The result could be the generalized collapse that has overtaken every earlier civilization. These are always accompanied, Tainter told me, by steep losses of technology, knowledge—and people. He is dubious of our ability to deliberately step down our complexity to stave that off.

All this reminds me of a tale told by the author Douglas Adams in one of his Hitchhiker's Guide books, *The Restaurant at the End of the Universe*. The planet Golgafrincham had too many people. So it contrived to keep the top professionals and the low-level practical workers, but rocketed all the middle-level "useless" people into space: security guards, for example, and telephone sanitizers. The remaining population subsequently lived happily—until they all died from a virulent disease contracted from an unsanitized telephone.

## CHAPTER 8

# The Pandemic That Never Should Have Happened—
# And How to Stop the Next One

> "We've got to dance with the virus.
> There's no choice."
>
> —George Gao, head, China Center for
> Disease Control and Prevention

In two campaign speeches while running for president, John F. Kennedy said, "When written in Chinese, the word 'crisis' is composed of two characters—one represents danger and one represents opportunity." Kennedy popularized the notion, and its use became widespread, including an appearance in Al Gore's Nobel Prize acceptance speech.

However, while it may have made for a good speech, it is not true. Apparently, this idea was the result of an optimistic mistranslation by Western missionaries in China in the 1930s. In fact, the first character does means danger, but the second one just means a time when things happen, or change.

Covid-19 has been, by anyone's reckoning, a crisis—and it's just getting started. Things are going to happen or change now, whether people take control of them in the broad interests of humanity or not. It may be an opportunity to achieve things we could not achieve before. The popularity of Kennedy's statement shows we recognize this deeper truth—that crises can provide those opportunities. Or, we might just be swept along by the economic and political storms the pandemic has unleashed and never deal with any of the underlying problems that got us here.

That would be tragic. In a minute, we'll look at our options. But first, let's look back at where we started and how we got to this point, where we can look toward the future. I called this the pandemic that never should have happened—and said we could possibly stop something like this from happening again. Let's see how that adds up.

In Chapter 1, we saw that Covid-19 started as a cluster of unexplained cases of pneumonia in Wuhan, China, with the first cases in November 2019. In late December, Chinese authorities told the WHO about it—but also said the virus did not spread person-to-person, although the doctors knew it did. With that as the official story, no large-scale containment efforts and public health messaging aimed at slowing spread of a contagious disease could be undertaken in Wuhan.

It's hard to imagine what local health officials, who ordered doctors to keep quiet as the epidemic grew, thought would happen. Maybe they thought they could keep most cases of the infection safely contained in hospitals. Everyone remembered SARS, another coronavirus, and people with SARS normally didn't spread the virus until they were quite ill.

Secrecy can also become an instinct in authoritarian systems, though, says writer and sociologist Zeynep Tufekci. As we saw in the chapter on SARS, China deemed outbreaks state secrets unless officials gave permission for them to be made public. This is not unique, or new: the International Health Regulations, now a cornerstone of global epidemic management, arose from efforts in the 1800s to stop governments from keeping cholera outbreaks secret and causing problems for shipping.

In early January 2020, officials in Wuhan obscured matters further by decreeing that someone could be tested for the new coronavirus only if they had been exposed either to the now-closed fish and wild animal market linked to many early cases or to a known case. As the virus spread in the population, increasing numbers of infections had no such links, so this guaranteed they would not test many cases. Europe did the same thing initially with swine flu in 2009, and many countries early in the pandemic refused to test people for Covid-19 if they had no direct link to China, even after the virus was known to be elsewhere.

Tufekci suspects that a culture of suppressing bad news and passing decision making up the hierarchy may have meant that President Xi Jinping didn't know how bad things were in early January—but in an authoritarian system, only he could change the story. Things were bad, though: by January 20th, there were so many cases in Wuhan and increasingly elsewhere that only drastic containment measures ahead of the Lunar New Year holiday could prevent the virus spreading out of control across China. Chinese scientists announced the virus was contagious, and the lockdowns began.

As we saw, research suggests that if those measures had been taken earlier, the epidemic might have been knocked back, although not knocked out. But would anyone have realized those measures were needed? The virus that causes Covid-19, we now know, can be difficult to stop by just isolating cases and tracing their contacts without added social distancing, even though that worked for SARS, as doctors in Wuhan would have known. Covid-19 is much easier to catch than SARS, and unlike SARS, it is spread by people without symptoms. The measures imposed after January 20th meant cities in China outside Hubei province, where Wuhan is located, never needed the total lockdown used in Hubei, but as the WHO reported, many of them found they needed social distancing as well as isolation and contact tracing to halt the epidemic. In early January, public health officials in Wuhan didn't know any of that.

Mathematical modeling shows that the kind of measures China eventually took in late January, including severe limitations on people's movements, would have vastly reduced the size of the epidemic if they had been applied in early January. But even if authorities had gone public about the virus being contagious then, it isn't clear that they would have imposed control measures that extreme, knowing only what they knew about Covid-19 at the time.

They would probably have done what worked for SARS, and it would not have been enough. Beside social distancing, widespread testing to catch pre-symptomatic or symptomless cases would also have been needed. It should be noted that, even knowing a lot more about the virus than China did at the outset, several countries, including the US and the UK, were slow to impose the testing, distancing, and containment required, and that did work—as demonstrated by countries that did impose these measures, like South Korea and New Zealand.

So it seems unlikely Covid-19 would have been stopped completely if China had publicized the full situation and taken more extensive public control measures earlier. But the spread of the virus, within China and out of it, might have been more limited and more controllable, especially if all this had happened before the end of December.

And if scientists and health authorities everywhere could have used this head start to tackle Covid-19 earlier and used viral sequences from China to test travelers from affected areas and look for cases at home, we might have headed off the eventual steep rise of infection in more places. But the virus still would have invaded poorer or less controlled countries that could not have taken these measures, and would have multiplied there, making it hard to prevent global spread. We would have needed an earlier massive shutdown in air travel to prevent that, which would have been unlikely.

Certainly there are what-if scenarios that suggest much earlier containment measures in China could have meant no pandemic at all, or maybe an epidemic in China and a few controlled outbreaks around the world. But the real clincher for me is looking at how many countries disregarded the WHO's advice on control measures even after it was clear how bad the virus could be. Even if they'd known earlier, I'm not sure how many would have done what was needed in time. Hindsight helps you win the next battle, not the last one.

Of course, because of this battle, we can at least hope that next time countries won't be as slow to see the danger they are in from a viral infection emerging anywhere. With luck, we have at least been jolted out of the blind complacency and outright denial about infectious disease that delayed most of the world's response to Covid-19.

*So, lesson for the future number 1*: we need a high-level, authoritative system bringing countries and international agencies together to collaborate on disease, so that no one conceals important details about worrying outbreaks and everyone works together from the beginning. At the very least, we need surveillance systems that will spot clusters of cases early, when an infectious pathogen might still be contained—that computerized alert system China installed after SARS, or something like it, in far more places would be a great start, especially if the alerts were shared widely. We will look at possible ways of doing that later.

Also, the world needs to start taking the threat of pandemics, and the warnings of its scientists, seriously. Covid-19 is taking care of the first part of that. As for scientists, that will always depend on how inconvenient their advice is versus how venal their government is. But we can at least hope listening to scientists will become more of a norm now that Covid-19 has shown us how desperately a modern society needs to rely on facts, evidence, and honesty, rather than secrecy, ideology, or wishful thinking.

Having looked at how Covid-19 emerged, in Chapter 2, we looked at emerging diseases generally. By the 1960s, we had largely defeated the old infectious diseases with prosperity and vaccines, whereupon we disinvested in the kind of public health needed for infectious disease, despite the wake-up call of AIDS in the 1980s, warnings of more new diseases from US scientists in 1992, and evidence by 2008 that we were contracting zoonoses from wildlife at an increasing rate. The WHO made a list of the most worrying pathogens, including coronaviruses and horrors like Ebola and Nipah, so we could make vaccines and diagnostic tests for them. Very few are ready yet.

I call Covid-19 the pandemic that never should have happened. Yet scientists have been warning, increasingly, about the growing risk of pandemics since 1992. How is it possible to warn that something is going to happen and then say it never should have happened?

Easy: that is the whole point of warning. We didn't act enough on the warnings, and there have been plenty. I wrote an article in 1995 entitled "Can We Afford Not to Track Deadly Viruses?" It was about a WHO plan to monitor emerging diseases in the wake of an Ebola outbreak in Central Africa—except WHO member states weren't going to approve enough funding. Could we have done better? We could certainly have improved our systems both for spotting emerging diseases and for responding to them. The willingness of countries to pay for that kind of surveillance and response increased somewhat after 1995, but not enough.

*Lesson for the future number 2*: Now is the time to improve our systems for monitoring and response, first by beefing up surveillance for emerging disease, and second by investing in drugs, vaccines, and diagnostics for the threats we already know exist. Let's not let the coronavirus take our eye off Nipah and the others. Especially Nipah.

Coronaviruses were on the WHO's list of worrying viruses because of what Covid-19's relative did 17 years before. Talk about being warned. In Chapter 3, we looked at three warning shots from previous outbreaks of coronaviruses: SARS, MERS, and, in pigs, SADS. SARS offered us two big lessons that nations clearly haven't learned: protect health care workers and immediately tell the world when a threatening new infection breaks out. There has been progress on that second lesson since 2003, but obviously not enough, given what happened with Covid-19 in China. Our defenses against these viruses have been stymied by problems of capitalism as well. Despite all the warnings, we didn't develop any remedies for coronaviruses because after SARS was stamped out there wasn't an obvious market. Producing medicine for the public good rather than profit may be coming back, and it's about time.

*Lesson for the future number 3*: There are two parts to this lesson. One is about laying in supplies of existing remedies. We need PPE, personal protective equipment, for health care workers—SARS should have taught us that, but now Covid-19 is forcibly reminding us. Besides serious stockpiles, we need surge capacity in manufacturing. If countries don't learn even that much after the toll Covid-19 has taken on nurses and doctors in so many countries, I despair. We all should.

The second is about developing new ones. Profit-driven markets can do wonderful things, but not everything. We need to stop relying on them to do what only governments can do and develop products we desperately need for the public good, including new antibiotics, vaccines that everyone can afford—and better ventilators, because respiratory viruses will always be among our biggest threats. The US tried to do that and failed, another time market forces trumped public good. There are dozens of calculations showing the cost equivalent in fighter jets or nukes, which governments apparently can afford, compared with the costs to develop, produce, and stockpile the lifesaving medical goods we need.

So, what about stopping these viruses at the source, or at least knowing enough about the sources to know what's coming? In Chapter 4, we looked at bats, why they have so many viruses, why killing them is a very bad idea, and how the Wuhan Institute of Virology not only found the exact viral gene sequences of SARS lurking in bats from one cave, but also found viruses that were very close to the one that later caused Covid-19. Meanwhile a lab in the US found that these viruses, straight from the bat, caused disease in mice primed with the human receptor protein ACE2 and had no trouble invading human cells.

In the scientific papers they published, the researchers issued very explicit warnings about the pandemic potential of these viruses. There appears to have been no action taken on these warnings, except that the US government research project that collaborated with the Wuhan Institute had its grant renewed—only to have it canceled again when, amid the hysteria of the pandemic, unsubstantiated allegations arose that Covid-19 actually escaped from the labs that tried to warn us about it.

Outside commentators with various ideological axes to grind have seized on the labs' warnings to blame the labs themselves for the pandemic, in a kind of molecular version of shooting the messenger. It's worth noting that those labs did this work for 15 years and more without any sign of a problem. Meanwhile the same species of bat as the one found carrying the virus lives in Hubei, there may have been a colony of live bats in the city, and bat feces is a widely used eye medicine. Surely all that seems a greater risk.

For now, *lesson for the future number 4*: when publicity-shy, certainty-averse scientists put these traits aside and start screaming that there's a really threatening thing out there, we need to listen and make it someone's job to respond. I have no doubt this lesson will eventually filter through, when climate change starts causing massive crop failures, uninhabitable cities, and unprecedented waves of refugees. By then of course it might be a little late to act on it.

But here it is, the ultimate reason why Covid-19 is the pandemic that never should have happened. We may or may not have been able to contain it once it jumped from bats to humans—*but it should never have jumped*. We knew enough 15 years ago to begin avoiding bats, and bat products, and bat anything that might transmit their wealth of viruses. And according to all the science we know now, the Covid-19 virus came from bats—not civets, not pangolins, not raccoon dogs, and certainly not snakes (that was never a scientifically valid claim). It came from bats, and so do a lot of other viruses.

But we need bats because the rest of our ecosystems depend on them, especially globally vital resources like rainforests, never mind our food crops. So, we should give bats plenty of space. We certainly shouldn't build livestock farms near their roosts, and perhaps we should give people who can't avoid encountering bats extra disease surveillance and health care to swiftly catch any virus that jumps. But it's actually not easy to catch a virus from a bat. We really had to work at this. Let's stop.

In connection with this, may I respectfully make a suggestion? Tradition is very important, and traditional medicine is often valuable, but perhaps using bat feces to treat eye disorders is one practice we might consider letting go of. This is not because it's feces—indeed, Western medicine is now learning uses for that long known in China—but because of what we now know about bat viruses. The Chinese people asking for that particular remedy to be taken out of the *Chinese Pharmacopoeia* and traditional Chinese medicine shops have a point. There are safer ways to get vitamin A.

I'm not sure why people assume that for Covid-19 to emerge, a bat virus had to be transmitted via an "intermediate" species or a laboratory escape, when a lot of people are using bat feces as medicine. Granted, many of the viruses present in the feces may disappear as it is dried, but do all of them, every time? And even if drying means users aren't at risk, the people who gather and process it are, and they transmit viruses like anyone else. The fact that at least some purveyors of Traditional Chinese Medicine online had stopped selling bat feces by May 2020 "because of Covid-19" suggests the risk is being recognized.

The people trying to at least clean up live animal, bushmeat, and wildlife markets everywhere, not just in China, also have a point. People in Africa who depend on fruit bats for protein pose a dilemma that could perhaps be addressed with respect and research. Chinese markets, whether or not one was the source of Covid-19, harbor other viruses, notably bird flu—we know because Chinese scientists find it there and have called for closing the markets because that has stopped outbreaks. Probably wet markets elsewhere should also be cleaned up. Yes, such markets have a long tradition, but those years of history didn't take place alongside modern intensive agriculture, megacities, or our hyperconnected world, which magnify the risks they pose of swapping pathogens among species, including humans. We can find ways to provide these goods safely.

Speaking of flu, in Chapter 5, we looked at the one virus we know is going to stage a pandemic, how it does this, and how the swine flu pandemic in 2009 precipitated an attack on the WHO that may have made it harder to react to Covid-19. Meanwhile, many countries struggled to respond to Covid-19 because their only pandemic planning was for flu, which requires a different response. We still need those plans (and more) though, if only for bird flu, which is highly lethal and might be able to go pandemic in people while maintaining a death rate that would make Covid-19 look like the common cold—although another relatively mild pandemic flu might be the next one to win the genetic lottery, and we need to respond to that too. WHO officials have suggested we might need a more nuanced response to different severities of pandemic flu, or other outbreaks judged to merit a WHO emergency declaration— although it's worth remembering that however a virus may start out, we can't predict how it will evolve.

For now, strife over the risks of lab work to explore the pandemic potential of bird flu viruses suggests we should monitor high-containment labs more closely, transparently, and internationally. As discussed earlier, a good example of how *not* to do that occurred in April of 2020, when US funding for research involving the bat coronavirus lab in Wuhan was summarily halted.

The lab's chief scientist, Zhengli Shi has said the genetic sequence of the virus that causes Covid-19 does not match any they have sequenced. Unsequenced virus from a bat sample, or an actual bat, could in theory have infected someone if there was a failure to apply the rules requiring stringent protection. But similar viruses have infected ordinary people living near bat colonies and, as we saw, could well infect people who collect, sell, and use bat feces. That seems to be by far the bigger risk.

Labs keep careful records. Their role, if any, in this can be investigated, as the WHO is proposing to do. We need a transparent, international system of inspection and accountability for these types of labs and open, international decision making about which research is, and is not, worth the risk, to ensure that important work is done, and done safely, and that labs are not randomly accused when disease emerges. We desperately need the science, more than ever, but when the risk is global, control and responsibility should be as well.

One thing we can say for sure: Covid-19 was not created in a lab. In an analysis in the prestigious journal *Nature Medicine* published in March 2020, scientists admitted that we simply would not have known enough to do it. We wouldn't even have guessed that the bit of protein Covid-19 uses to attach to human cells would work so well. It does, though.

So, *lesson for the future number 5*: a flu pandemic is coming. By now, it should surprise no one to hear that we aren't ready for a bad one. We've already done a lot of the homework on pandemic preparedness for flu, though, and those plans should now be revised in light of the hard lessons we are learning about pandemics from

Covid-19—and actually be made ready to roll. At the same time, the global framework the WHO runs to monitor flu evolution should not only be maintained, but more generously funded and expanded to include other worrying virus families. An international collaborative effort among scientists who study the evolution of pathogens and livestock scientists—who are now rarely even in the same room—should aim to wean farm animals off vaccines or other management practices that foster dangerous pathogens. We desperately need ways to make flu vaccines for people much, much more quickly—and, if possible, we need a universal flu vaccine as well. Yes that's right, both kinds of vaccine, belt and braces. Flu deserves it.

After our trip through the long record of largely disregarded warnings that led to this moment, we looked in Chapter 6 at what we should be doing to prevent the next pandemic—whatever it is—or to respond and contain it fast if one starts. We need decent pandemic plans. We need stockpiles of response equipment. We need worldwide surveillance for emerging disease, as much as possible by local experts who understand their own situation, but have a global network of colleagues and resources at their backs. We need a lot more basic work on diagnostic, vaccine, and drug technologies, and we need to deploy the capabilities we have so we are ready to use them, fast, and everywhere. It sounds expensive, but as we are learning, it will almost certainly cost less than the next pandemic—if we can still muster the organization, and the cash, after this one to take the precautions we now know we need.

*Lesson for the future number 6*: we need to hold governments accountable for their promises, now, to do all this. Actually, this lesson is one we should act on now. The G20 group of the world's richest countries promised to take action on pandemics in late March 2020, including holding a joint meeting of finance and health ministers "in the coming months" to create "a universal, efficient, sustained funding and coordination platform to accelerate the development and delivery of vaccines, diagnostics and treatments." That meeting may still be in the future as you read this, or it may be long gone. But two-thirds of all the people in the world live in G20 countries—so more likely than not, dear reader, one of the governments that made that promise was yours. Whatever happens or happened with that meeting, the participants need to be held to account. Try and do that.

If we don't hold those responsible accountable now, we might see something like our trip to the dark side in Chapter 7. Few people realize that the rapidly increasing complexity of our globalized society is increasing risk in ways that could be catastrophic. That is how the outbreak of a new respiratory virus in China rapidly went pandemic and why painful economic domino effects just from limiting our interactions to slow spread of the virus have propagated globally. We looked at the possibility of a much worse pandemic, with a higher death rate, and discovered that the widespread belief that diseases that go pandemic always become milder is a myth. In fact, if we aren't careful, vaccination for Covid-19 could make that virus worse. I looked at pathogen evolution and complex systems together, not just because both are frightening, but because it's the two together that pose the real threat: a severe pandemic could precipitate cascading failure in our complex global support systems. Especially if the low-income people who hold a lot of it together are further weakened by growing economic inequality.

*Lesson for the future number 7*: pandemics are serious. After weathering the initial onslaught of Covid-19, we can't go back to normal. Normal is what led to this, and more of it means there will be more pandemics, and they could well be worse. We have to take the obvious preventive measures of Chapter 6: stockpile PPE, build vaccine plants, do more disease surveillance, and plan. But the possibility that a big disease event will trigger epidemics of collapse throughout our global systems—food, water, security, financial, even nuclear—is the bigger problem we must try to fix, because that tightly linked complexity is why the risk, both of having pandemics and of their impacts, is increasing.

We must manage our global system with some understanding of how complexity works, taking advantage of the global shock caused by this pandemic to build looser connections, less efficiency, more redundancy, and resilience into global supply chains, economies, and governing structures, even if that is never the cheapest option. If a few connections collapse here and there, complex systems experts suggest it might be more opportunity than disaster: "creative destruction" might let new, more resilient patterns emerge, especially if we rebuild with that in mind.

We must grasp that a much worse pandemic can happen, and it could trigger nonlinear effects in our global system that could lead to collapse of local systems or global ones. Some of the world's smartest scientists say that is what we risk. Every disaster movie starts with someone ignoring a scientist.

So here we are. Are we back from the dark side now? Is there any good news?

Yes. Crisis can be an opportunity, even if that is badly translated Chinese. We desperately need to redesign the systems that failed to contain this pandemic if we are to, with luck, prevent or at least contain the next one.

If you take one thing away from all we have looked at in this book so far, it should be that people have been predicting this pandemic for decades, yet we were not prepared. Covid-19 was an unnecessary catastrophe: we knew enough to keep people away from bat viruses, to develop drugs and vaccines for coronaviruses, and to set up transparent, truly global surveillance networks for outbreaks of potentially pandemic disease. Such surveillance systems would mean that if an outbreak occurs despite prevention efforts, rapid, aggressive action can be taken to at least try to limit its spread.

As Covid-19 emerged, one local bureaucracy delayed the warning—and there was no international agency that could go in and verify what was happening on the ground, immediately, on behalf of everyone else. Then we didn't have the global public health infrastructure to ensure every country's response was adequate, even though inadequate response in any country could mean increased infection in others. We didn't have crisis-management systems that could try to counteract local or national governments' denial and delay—even though that affected everyone.

The WHO did for Covid-19 a bit more than what in 2013 I predicted it would do for H7N9: it issued advice, held daily briefings, organized R&D, and got PPE and test kits to poorer countries. Which is to say, it did what it could. Yet many countries were still mired for weeks in inaction until the disease hit. "The virus is faster than our bureaucracy," admitted Italian authorities, when what was needed was more like mobilizing for war.

So how do we fix that? Obviously, there must now be major investment in the scientific preparations we should have made for this pandemic. Jeremy Farrar, head of the Wellcome Trust in England, says countries need to invest in public health and in the clinical, social, and basic science of infectious disease. "You will need it," he said in a talk to the US National Academy of Sciences in April. That includes many countries' long-atrophied capabilities to do basic epidemiological controls: isolation, quarantine, and contact tracing.

Making our response and alert systems truly global might be less obvious. As I argued, we need a high-level, authoritative system bringing countries together to collaborate on disease, so that no one conceals important details about worrying outbreaks and everyone works together. Easier said than done, perhaps, but where can we start?

Many criticize the WHO, although frankly I think it just makes for an easy target. There were certainly things it might have done better this time: I think it could have called Covid-19 a public health emergency, and then admitted it was a pandemic, earlier, communicating the real urgency of the situation rather than holding back for fear of scaring people—or perhaps offending governments. But the WHO has few choices in this regard: it can do very little independently of its member states. Yet it remains the world's only global health agency, and it must be part of building a better system.

First, let's look at why we have to organize globally at all.

Globalization has become a bad word in some circles. Indeed, in Chapter 7, we looked at how a lot of our vulnerability in a pandemic is due to our tightly interlinked global systems. But the bad part of that is not the "global" part. It is the part where it is all so tightly and efficiently linked. That optimizes profits, but it also creates a rigid network that transmits shocks. In this pandemic, closed clothing stores in Europe created unemployment in Bangladesh, while factory shutdowns in China threatened the availability of electronics and essential drugs in the US. Some experts think the tightly coupled, fragile global financial system, barely ten years removed from its last worldwide crisis, has already come close to meltdown too.

But is the answer to make unemployment in Bangladesh permanent by bringing those clothing factories "home" or to shut down the amicable global trade links between great powers that have fostered the longest stretch of relative peace the world has ever known? If Covid-19 teaches us anything, it is that we really are all in this together.

Some people in the anti-globalist, or just plain nationalist camp strongly believe we should not be organizing ourselves on a planetary scale at all. Yet, given that virtually all our economic and cultural activity is now on that scale, it is hard to argue we should not also be managing our affairs on that level. Just having eight billion people filling virtually every available niche on this planet makes us global whether we like it or not. We can no longer run our affairs in isolated groups, while even a fraction of us might do things that affect everyone: besides disease, there are greenhouse gases, ozone-depleting chemicals, overfishing, financial instability, pollution, deforestation, cybersecurity, nuclear weapons—the list goes on. To even try to get ahead of the cascading failures that can result—like pandemics—we have no choice but to organize on a global scale as well. If the pandemic doesn't teach us that—well, global warming will, but possibly not until it's too late for lessons.

"We've created a tightly knit socio-ecological system that reaches into all corners of the planet," says complexity expert Thomas Homer-Dixon. "If we're to grasp the nature of today's emerging global dangers and adequately mobilize ourselves to do something about them, 'we' needs to come to mean, to a lot of people most of the time, the entire human species."

So how do we do that, at least as it applies to preventing pandemics? Certainly, part of the answer must be strengthening the WHO so it can do the job we are already asking of it, acting as a kind of global civil service for health, if not a real political authority. The nation-states that hold most of the power in the world seem unlikely to permit an international agency to wield very much power of its own.

But we could at least allow it enough power and resources to play the supporting role effectively. As it stands, the WHO, the world's only organization charged with stopping pandemics and pursuing all other aspects of health that have an international dimension, has a budget of $2.4 billion per year for 2020 and 2021, virtually no real increase from the previous four years—and that was after a 20 percent cut in 2011 due to the financial crisis, with emergency and epidemic funding cut still further.

Yet over the past six years, the WHO acquired an emergency response capability, expanded its work on antibiotic resistance and the health threats of climate change, almost completed polio eradication, and is now leading the world's response to the pandemic. It was working on a shoestring as it was. With no increase in funding in that time, it is stretched pretty thin.

Then in April 2020, Donald Trump threatened to withdraw US funding from the WHO, which is 15 percent of the agency's regular funds. Larry Gostin, an expert in public health law, called it an effort to deflect blame for the slow US response to the pandemic, even though the WHO had been screaming for weeks that countries needed to do more. Gostin charged that the WHO's budget is a third that of the US CDC, and the CDC doesn't have to respond to health emergencies across the planet. The WHO also runs a plethora of programs to strengthen health systems in poor countries, which, we should all now realize, benefits all of us.

Worse, it gets to spend only a third of its budget as it likes; the rest is earmarked by member states for pet projects. Its emergency fund is run on voluntary contributions, and after using most of that to contain an Ebola outbreak in the Democratic Republic of the Congo between 2018 and 2020, it had a risible $9 million left to help poor countries respond to Covid-19. It took countries weeks to respond to its emergency appeal for pandemic funds. Gostin says that to act in the interests of the world, the WHO needs a doubling of regular funding, and that needs to be less subject to the partisan interests of the richest member states.

Mostly, though, Gostin says, "We need to finally recognize that this novel coronavirus is the common enemy, and unite as a global force to overcome it." In May 2020, UN Secretary-General António Guterres echoed this, saying the virus spread out of control because "the world was not able to come together and to face Covid-19 in an articulated, coordinated way." The astronomical costs of the pandemic might finally ram home the idea that events with potentially catastrophic global effects should be a shared, global responsibility, and not be subject to the interests—or just the local bureaucratic limitations—of any one country.

How do we come together, as Guterres says, in a coordinated way if the UN and the WHO have not done that already? Right now, most power rests with sovereign nation-states, especially the 20 or so rich, powerful ones. We saw how national sovereignty finally yielded to global health security with SARS. But the WHO is still very much the creature of its 194 member states. When a country's interests do not coincide with those of the world at large, the WHO represents the interests of the world, but the country can often win.

The most obvious example is China's insistence in early January that Covid-19 was not contagious. But you don't have to be big and powerful: in 2014, the WHO's response to the Ebola epidemic in West Africa was delayed when the government of Guinea was reluctant to report true case numbers for fear of discouraging foreign investment.

I would like to suggest two kinds of solutions. Neither one involves replacing the WHO—as I said earlier, it is the only game in town. In fact we need a stronger WHO that can act in the interests of the world despite occasional conflicts between the world's interests and the claims or abilities of nation states. How do we do that?

One way is to start from the recognition that nation states seem unlikely to give an international agency the power it needs to overrule nation states. So if countries have all the power, they have to find some way to use it in the common interest. If the WHO cannot tell a big country what to do, other countries will have to. Which might work, especially as they are all trading partners.

A global government, as governments are usually understood, is not likely to work. Complexity scientist Yaneer Bar-Yam says that when social systems get too complex, old-fashioned hierarchies, with one guy (it's usually a guy) in charge, don't work anymore because one person can't get their head around everything. Hierarchies are already devolving into global networks as the real power structures in many areas of global concern, writes author and governance expert Anne-Marie Slaughter, especially things that can be managed by networks of experts.

So we need a network. We now have an annual meeting at which WHO member states tell the WHO what they want it to do. What if we also had a more constantly convened, high-level council of countries to deal with global threats, that could demand that individual governments act on big deals like suspicions of incipient pandemics and other problems that could have impacts far beyond one country? And what if it was advised by a network of scientists-on-call, like the one the WHO now convenes for emergencies?

Bill Gates also wants a broader structure for managing disease, and envisages a globally networked approach, with a mix of national, regional and global organizations all focused on pandemic prevention, playing "germ games" the way armies play war games, to hone their skills at spotting and containing simulated outbreaks. Perhaps that would foster more openness among countries.

It shouldn't be beyond our wit to design something that works in the common interest, as that, by its very definition, benefits everyone. The current dispensation based on the unassailable sovereign rights of nation-states (especially rich ones) in any conceivable situation does not work in a world of shared catastrophic risk. The world is networked, and it takes a network to run a network.

There is a second kind of solution we might consider. When you talk to veterans of international agencies, they will tell you that countries are never going to give up sovereign power to any international agency. For proof, look no farther than the 2005 revision of the International Health Regulations. It gave the WHO the power to ask a country about an outbreak of any disease that might be an international threat, which it had heard about from other sources. Before that, it could only ask about a few diseases and only if it had been told about it by the government concerned. The revision also allows the WHO to talk about an outbreak publicly—if the outbreak is already public knowledge. That's it. And it took until 2005, after SARS had almost spun out of control, to get that much of a concession. And even that took hard negotiating.

But the WHO still can't investigate an outbreak directly unless the country invites it. It could not investigate China's claim that Covid-19 did not spread person-to-person before Beijing admitted it did on January 20th, 2020, and it could not actually go to China to investigate Covid-19 until February.

It's not as if it didn't have an inkling of what was going on. "I was concerned that there had been no report of further cases or any information about transmission, particularly information about possible human-to-human transmission, between the announcement of the outbreak by Wuhan authorities on December 31st and January 17th," says John Mackenzie of Curtin University in Perth,

Australia, then a member of the WHO's emergency committee on Covid-19. Yet WHO could not investigate without an invitation from Beijing and Beijing's approval of all the mission members.

This will not change unless WHO member states approve a wholly new treaty, or perhaps an add-on protocol to the IHR. The IHR is legally binding, and it requires countries to improve their surveillance and public health capability and to assess their own ability to detect and respond to outbreaks, says David Heymann— but like all international treaties, it has no way to enforce this. But enforcement is not how treaties get things done: verification is. There are already treaties in which countries have set aside one tightly circumscribed area of national sovereignty in the name of global security. We have one governing nuclear material, another on chemical weapons, and another on the ozone layer.

Members of the Nuclear Non-Proliferation Treaty must declare any uranium or plutonium that can be used for nuclear weapons, prove they haven't diverted any for weapons, and submit that to verification inspections by the International Atomic Energy Agency, or IAEA. The IAEA caught Iran cheating twice and imposed an inspections regime that was keeping it from enriching too much uranium—until Donald Trump torpedoed the agreement in 2018. The five official nuclear powers have not yet given up their weapons as promised in the treaty, and four countries have acquired them despite it, but, weapons experts tell me, the world is nowhere near as awash in nuclear material and weapons as was looking likely when the treaty came into force in 1970.

The 1997 Chemical Weapons Convention (CWC) bans making or stockpiling a list of known weapons, such as nerve gas, and prohibits using any chemical as a weapon. Member states—everyone but Israel, Egypt, North Korea, and South Sudan—declare any facilities they have that could make these things, and inspectors from the treaty's Organisation for the Prohibition of Chemical Weapons (OPCW) verify they aren't, and also check ordinary chemical plants. There are holes in the verification regime but it has generally worked, although the chemical disarming of Syria in 2013 may be weakening. A similar treaty banning biological weapons was also supposed to be equipped with a verification protocol, mandating inspections of biolabs. Although the convention still stands, it is largely toothless without that protocol, which was torpedoed by the US in 2001.

The CWC's real innovation is that someone can charge a member state with not declaring a chemical weapon, or using one illegally, and ask for a surprise "challenge" inspection. Treaty countries have all agreed to "anytime, anywhere" inspections with no right of refusal, except the US, which passed a law allowing it to refuse. No one has ever demanded a challenge inspection, although the OPCW's destruction of Syrian chemical weapons in 2013–2014 was one in all but name. In another kind of check on bad behavior, the 1987 Montreal Protocol to the treaty banning chemicals that destroy the earth's protective ozone layer allows member states to slap trade sanctions on countries that break it. It never has, but at one time at least, we all agreed such a threat was appropriate.

Although these treaties have had rather spotted records, they have if nothing else established what weapons experts call a "norm" against these weapons and chemicals: we have all agreed that we're not supposed to have them. "It somehow has to become an international norm that you don't just let infectious diseases fester," without fully reporting them, says Amesh Adalja at Johns Hopkins.

We already have a treaty requiring countries to declare any worrying outbreaks of disease, the International Health Regulations, but verification could make it really effective. We even have a ready-made verification agency: the WHO. Verifying outbreaks would have a completely different dynamic from a weapons treaty. After all, a country with banned weapons presumably acquired them on purpose and plans to use them, if only as a threat. A country can harbor a disease just through bad luck or difficult geography. And it usually isn't intending to launch it at an enemy: the virus will arrive in other countries anyway, on the next passenger flight. Inspectors in these circumstances are friends, not adversaries.

On the model of weapons treaties, the IHR could require countries to make declarations about infectious diseases on their territory, regularly if things are normal, as a matter of urgency if there is a worrying outbreak. Then inspections could enable some kind of verification of what a country has said about its outbreaks, including that there are none. To believe a country's declaration that it had no worrying disease, we would need to know its local systems were able to spot one if there was.

That means verifying a country's surveillance systems in peacetime. There is already a similar type of verification in the program for polio eradication, in which the WHO participates: if a country says it hasn't found any cases of paralysis that turned out to be polio, it has to have found the number of cases of paralysis that weren't polio that you would normally expect in its population, so we know it was looking hard enough. This kind of system could finally enable poor countries, with help from rich ones, to develop disease surveillance systems we could all rely on.

In 2004, after China admitted it had H5N1 bird flu all over its territory, I wrote for *New Scientist* that we should "start controlling viruses the way we control nuclear weapons or ozone-depleting chemicals." The stakes, if anything, were even higher: were this flu to go pandemic, I wrote, "the economic cost, political toll and loss of human life would be colossal. Bring on the Pathogens Treaty."

Years later, I am even more convinced that some agreement like this must be the answer—and perhaps the wreckage of Covid-19 will make it politically feasible. Pandemic disease, as Covid-19 has abundantly demonstrated, is more devastating to more countries for a longer time than any chemical weapon could ever be, yet countries have agreed to inspections "anytime, anywhere" to prove they don't have chemical weapons and to make sure their neighbors don't. A pandemic actually can start anytime, anywhere—surely inspections that can keep pace with that are the only defense any country can really trust.

I ran the idea past a few weapons experts. They feel the world has grown tired of treaties, that "multilateral" cooperation among countries is no longer in vogue. Well, we're having a multilateral pandemic. Vogues change. Abstract concepts of national sovereignty might motivate treaty negotiators in conference rooms in Geneva, but in practice, with an unknown disease threatening, no country wants to look unreasonable to their partners in a global market over something that could threaten everyone. If some country had an outbreak, and the WHO asked to come investigate, and that country said no, how would that look?

Such an arrangement might also help us head off the inevitable blame games governments play regarding disease. It would give a country a chance to defend itself against charges that a pathogen of concern escaped from a lab or, from the point of view of the country's neighbors, a chance to verify whether or not it did. And the prospect of inspections might make such escapes less likely. In Chapter 6, we looked at risky pathogen research that might be made safer if it came under transparent, international supervision. That could be part of this treaty.

Besides, treaties are not all about carrying a big stick. They are also loaded with carrots, like the pledge in the IHR for rich countries to help poor ones monitor disease, and pledges in the nuclear, biological, and chemical weapons treaties for the rich to help poorer countries use those technologies peacefully. In these weapons treaties, treaty members conduct confidence-building exercises, where experts from other countries visit your installations and you visit theirs. And unlike governments, experts communicate easily, so cultures of secrecy are less likely: the virologist in Saudi Arabia who first encountered MERS solved the problem quickly by turning to a Dutch virologist, who in turn supplied the virus to qualified labs elsewhere for research or diagnostics. Everyone benefitted. There should be some international arrangement that fosters such exchanges, makes them safe and transparent, fairly assigns any patent rights, and makes any government objections look dangerously old-fashioned.

The need for international solidarity on disease is growing just with our existing pandemics: by 2030, Peter Piot, formerly the head of UNAIDS, told me, Zambia will need 3 percent of its GDP just to deal with HIV, and it will need help. Moral considerations aside, why should rich countries care about HIV in Zambia? For the same reason they care about whether another Covid-19 might be brewing somewhere: less disease means less poverty, and that in turn means less risk of emerging disease, as more prosperous people stop having to take health risks just to live.

In fact, we can't talk about solving the problem of pandemics without talking about global inequality generally. Covid-19 came from China, which is not a poor country, but the same cannot be said of areas prone to other worrying viruses, from Ebola to Nipah to something unknown now because it lives in a country with no virology or disease surveillance going on.

UN Secretary-General António Guterres said, at the G20 meeting that pledged pandemic preparedness measures in March 2020, "We must work together now to set the stage for a recovery that builds a more sustainable, inclusive and equitable economy." British columnist Tim Walker of *The New European* newspaper was hoping for much the same when he tweeted, "When this is over, we may have got used to better air, seen the point to international co-operation, that people don't have to sleep on the streets…We might twig there's more to life than nationalism and the economy. It could be a new beginning."

Jonathan Weigel and his colleagues at the London School of Economics call for a global solidarity fund for pandemic response and recovery in poor countries. "The developed world cannot heal if the rest of the world is in critical care," they write. "Renewed commitment to multilateralism and global solidarity is the safest path forward—for all of us."

Be it a global fund or a treaty pledging to monitor disease—and verify it—the imperative is clear, both when it comes to defeating Covid-19 and ensuring a human future less threatened by disease. We really are all in this together, and we'd better start acting like it. And ideally, make it legally binding.

Unfortunately, thinking like that, involving international cooperation, may soon be less common. Besides our biological immune system, psychological research has recently found evidence that humans have a "behavioral immune system"—a tendency to avoid people who may be carrying disease. Besides avoiding obviously ill people, psychologists think humans universally tend to conform to our own "in-group" and avoid people who are different from us, because originally, we were trying to avoid infection.

This was actually a risk during our early evolution. When we were wandering hunter-gatherers and encountered another wandering tribe, the strangers might have encountered different diseases and be carrying germs to which they had acquired resistance, but we had not. This was especially true because some aspects of disease resistance are genetic, and we would have shared fewer genes with another wandering tribe than we do now with fellow city dwellers. The disease risks that separated populations may pose to each other were confirmed with a vengeance when most of the native people of the Americas died of European diseases after Columbus arrived. Europeans got syphilis back.

There is evidence the behavioral immune system underlies tribalism and xenophobia. People with stronger disgust responses to descriptions and pictures of things that might pose a disease risk, like dead cats or rotten food, tend to be more xenophobic and politically conservative, as are people from places with more pathogens, now or historically.

Researchers have focused on "authoritarian personality," a set of personality characteristics, partly determined by genes, which include a desire for order, obedience, conformity, and cohesion within the in-group with which the person identifies. Having those characteristics made it more likely that a person would vote for Donald Trump in the US, or Brexit in Britain, in 2016, more than any other variable measured.

Cambridge psychologist Leor Zmigrod has discovered that people who live in US states and cities with a higher prevalence of diseases you catch from humans—but not diseases you get from animals, like Lyme disease—are more likely to have authoritarian personalities and to have voted for Donald Trump. States with more pathogens also tended to have more laws that restrict minorities, such as LGBTQ people. No other variable, such as education or life expectancy, correlated as well.

Other research has found that activating the behavioral immune system, either with a real disease outbreak or with disgusting images or mentions of disease, shifts people's political attitudes in an authoritarian direction. Canadian psychologist Mark Schaller, who coined the term "behavioral immune system," found that in 2014, Americans were more likely to tell pollsters they would vote Republican after the appearance of Ebola cases in the US, especially in places with a heightened interest in the disease as reflected by Google searches for "Ebola." The same went for conservative voting intentions among Canadians.

This fits with a history of sometimes violent xenophobia and hostility to strangers after epidemics. European cities slaughtered Jews and gypsies during the Black Death of 1347. In 1793, Philadelphia blamed a yellow fever outbreak on traveling actors. White North Americans have blamed cholera on Irish immigrants, AIDS on Haitians, plague on Chinese immigrants—Honolulu burned its Chinatown—and SARS and Covid-19 on ethnic Chinese. The Centre for Economic Policy Research says the 1918 flu pandemic led to greater mistrust of governments. As a presidential candidate, Donald Trump blamed Latin American immigrants for "tremendous infectious disease." All the claims were groundless.

A lot of psychological researchers are now collecting data on the political impact of Covid-19, Schaller told me. "If Covid-19 elevates the allure of authoritarian ideologies, the effects could be long-lasting," says Zmigrod, as she finds those ideologies are more common in places that merely had more infectious disease in the past than in places that didn't. That could be especially true if Covid-19 doesn't quite leave but continues to circulate.

This impulse toward division is troubling when the world needs greater collaboration, not less, to defeat the shared risk of disease. The least likely prospect for cooperation now seems to be between the US and China, with their respective leaders trading barbs and blaming the other for the virus. Yet the need is great. In February, Shi Zhengli, Kevin Olival, and 21 other emerging disease researchers made a detailed case for the US and China to work "synergistically" on research into pandemic threats. Only better understanding of disease ecology, they wrote, "can avert the increasing numbers of catastrophes in waiting."

The two countries between them dominate livestock production and the global trade in wild mammals, two major pathogen sources. China is the world's largest maker and consumer of antibiotics, more than half used in animals, hence a major source of resistant bacteria. The world's two biggest economies have a moral responsibility as "major drivers of the ecological change responsible for the emergence of new disease," the scientists argued—and, they noted, they also happen to have the world's biggest combined infrastructure for infectious disease research. Yet increased collaboration between the two could become less likely if disease really does promote authoritarian and xenophobic tendencies.

Optimists, however, are hoping that the shared threat, anxiety, and hardships many of us are experiencing will outweigh atavistic fears of infection and breed social solidarity instead of hyperactive behavioral immune systems. American author Rebecca Solnit has documented that, in the wake of many disasters, survivors support each other with generosity, resourcefulness, and altruism. I find myself repeating the catchphrase "we are all in this together" in descriptions of the pandemic, because events have overwhelmingly shown the truth of that, for better and worse.

The pandemic "could help catalyze an urgently needed tipping event in humanity's collective moral values, priorities and sense of self and community. It could remind us of our common fate on a small, crowded planet," hopes Homer-Dixon. "We won't address this challenge effectively if we retreat into our tribal identities. Covid-19 is a collective problem that requires global collective action—just like climate change."

Whether renewed xenophobia or recognition of our shared

peril dominates the world's response to the pandemic may depend on how countries deal with one thing: the apparently incontrovertible fact that Covid-19 started in China. Both the US and China have accused the other of originating the virus. Some American companies have launched lawsuits against China for covering up details about the disease in December and January.

Not everyone sees this in adversarial terms. In April, 101 top US scholars and former officials, including such senior figures as Madeleine Albright and Susan Rice, petitioned the US government to cooperate with China in fighting Covid-19. "China has much to answer for in its response to the coronavirus: its initial coverup, its continuing lack of transparency," they wrote. "Notwithstanding this, we the undersigned believe that the logic for cooperation is compelling."

In May, Ursula von der Leyen, president of the European Commission, called for an international independent inquiry into the origin of the virus, not with a view to assigning blame, but so the world can work together to prevent it from happening again. "It's in our own interest, of every country, that we are better prepared the next time," she insisted, calling as well for a "transparent" early warning system. "The whole world has to contribute to that." Her message: we need China to be part of this effort, and blame games will not help.

While it is true that China hid details of the virus from the world for a crucial few weeks, it is also true that China itself has sustained enormous economic damage, which as in other countries was caused more by its efforts to stop the spread of the virus than by the virus itself, although the death toll has also been horrible. Nor was it the only country that was slow in recognizing and responding to Covid-19. Lots of countries made mistakes, and we probably haven't made them all yet. Recognition of all that by all sides, including by China, might be a good place to start.

Pathogens come from all over. The last flu pandemic started on an American-owned farm in Mexico, the biggest-ever Ebola epidemic began with the infection of a two-year-old child in one of the poorest countries in Africa. The HIV pandemic was seeded in an African society upended by European colonialism. The Zika virus started in Africa and then traveled via Asia, Micronesia, and Polynesia to Brazil, then wherever in the Americas it could find the right mosquitoes, which were themselves transported worldwide by numerous countries. Of the viruses that are still mere threats, Nipah started in Malaysia and the very similar Hendra in Australia. This is a planetary problem.

Jeremy Farrar seemed to be addressing those very concerns when he spoke at a virtual meeting organized in April by the US National Academy of Sciences. "Throughout history the world has faced great crises. In the aftermath there is always a choice," he said. "Do we apportion blame, exact reparations and become ever more polarised? Or do we come together, learn lessons, make changes and refashion a more collective, cohesive world?"

Viruses don't care about human borders, identities, or ideologies— just human cells. The question now is: Do we care enough about defeating them to truly join forces?

## ACKNOWLEDGMENTS

This is what the book trade calls a "crash" book. Those are written in a very short time at junctures when a lot of people very much want to know about a certain issue. They are written by people who just happen to know a bit about the topic and are poised to go.

That's the position I found myself in with Covid-19. I have spent the past 36 years as a journalist for *New Scientist*, a weekly scientific magazine based in London. Since the late 1980s, a large part of my beat has been infectious disease, the kind caused by germs rather than toxic chemicals or faulty genes—including viruses like Covid-19.

Obviously, writing a book in two months is not the way I always pictured my first outing as an author. Inevitably, there are going to be rougher edges than might have been the case if I had had the more traditional year or so to do this. But I couldn't pass up a chance to tell people what I have been hearing for years about emerging disease, just when people are really able to hear it. To colleagues who say I'm crazy, you have a point. I still think we all need to take a step back and look at the big picture.

## Acknowledgments

There are many vitally important aspects of this pandemic I could not write about properly in a project like this. Readers may be angry that I do not go into damning detail about this or that politician who screwed up their country's response and thereby caused deaths that might have been avoided. Yes, there have been those. Holding them to account will be massively important as the dust settles from this first wave of the pandemic, if it gets a chance to. I couldn't do an analysis, certainly not one that would have been appropriate for a book, of events that are still the stuff of news and changing by the hour. I'm sure many, many of my colleagues will do that, brilliantly, elsewhere.

I also couldn't tell you what drugs and vaccines or economic and social remedies work, and why. I could start to tell you how some governments contained the virus while even more failed to, but I could not tell you yet how that plays out as the virus comes back for repeated rounds—or even if it will. Those analyses will be happening, probably, for the rest of our lives and beyond.

What I could tell you is why we already knew this was going to happen and how the predictions panned out. I could tell you where we have got the wrong end of various sticks about pandemics. I could try to convey the big picture on the pandemic risks we run and, most importantly, what we should do, now, to try to stop this—and possibly worse—from happening again.

Obviously, I could not be writing this book if I hadn't been privileged to write about these subjects and allied ones for decades for a magazine as uniquely dedicated to the juncture between science and society as *New Scientist*. I could talk to the scientists and get that big picture because of the respect most of them hold for the magazine.

But getting the story out meant putting a lot of editors through some stressful moments. There were the features that took far too long as I dove down rabbit holes of unexpected twists in stories ranging from the impact of megadeath to the disappearance of Indian vultures. There were the countless news stories, week after week on harried deadline days, as news editors and subs fielded everything from last-minute shifts in the story as new facts emerged, to correctly spelling the names of obscure viruses and virologists, to occasionally standing up to a boss's cries of "not another story about flu!"

There have been far too many editorial survivors of those deadline days to single any out by name, for fear of unfairly neglecting others. Apologies. But you know who you are. Thank you.

I would like to single out the editor who got me started as a news journalist, though. Back in the 1980s, I was the survivor of a bruising decade of grad schools and labs, and I had decided to write loftily about science for the masses instead.

Then Fred Pearce, the new news editor at *New Scientist*, heard I was based on the Continent and started sending me off to do news stories. I rapidly became hooked, and I'm grateful. Obviously, some of the information in this book was learned or unearthed in the course of investigations I did for *New Scientist* over the years, and I hereby totally acknowledge that. I note in the text when I am particularly indebted to a particular story.

I would especially like to thank the various scientists and allied experts who were patient enough to explain complicated bits of their life's work to me, at length and repeatedly, so I could write those stories—often as I, or they, were facing a screaming deadline. There have only been three I know of who refused to ever speak to me again after I wrote the story. And I don't regret those.

I would also like to thank the scientists who helped me with all the new science I had to digest to write this book, even though most of them are putting in long shifts and hazard duty on the front lines of Covid-19. All the mistakes that got through are, obviously, mine. I'm sure my scientist contacts will be letting me know about them.

I would like to express unlimited gratitude to my agent, Max Edwards, whose wild harebrained scheme this was—if they ever let us go to restaurants again, Max, I owe you serious lunch. I would also like to express my greatest respect and gratitude to my editor, Sam Raim, who took on the unlikely task of trying to get a book together and presentable in a ridiculously short time, all while working from home. The mistakes that remain despite his desperate ministrations are mine. Except for all the Oxford commas: those are his.

Finally, like all authors, I need to thank my long-suffering family who had to deal with me disappearing into my office for weeks and muttering obsessively about disease when I emerged—and that was for the past several decades, although it admittedly got a lot worse during the frantic weeks I was writing this book. My family has been endlessly supportive, despite being locked down in either various places as Covid-19 raged. Thank you to my husband for constant cups of tea, watering the roses, and holding everything together. Thanks as well to both him and my daughters for playing the critical reading public with a few of these chapters, all while doing their jobs from lockdown—and in the case of my daughters, recovering from their own encounters with Covid-19. When you get your sense of taste back, I promise you carrot cakes till the cows come home.

# NOTES

Unless clearly indicated in the text, where a source has not been provided for direct or indirect quotation, the material comes from an interview with the author.

## PREFACE

xii **As it stands...watch helplessly:** Debora MacKenzie, "Why we are sitting ducks for China's bird flu," *New Scientist*, May 1, 2013, www .newscientist.com/article/mg21829150-200-why-we-are-sitting-ducks -for-chinas-bird-flu.

xiii **As far back...coming years:** Institute of Medicine (US) Committee on Emerging Microbial Threats to Health, *Emerging Infections: Microbial Threats to Health in the United States*, eds. Joshua Lederberg, Robert E. Shope, and Stanley C. Oaks, Jr. (Washington, DC: National Academies Press, 1992), doi.org/10.17226/2008.

## CHAPTER 1

2 **On the evening...committee:** ProMED-mail, "Undiagnosed pneumonia—China (HU): RFI," *ProMED-mail Archive 20191230.6864153*, December 30, 2019. Available at: www.promedmail.org. (Brackets are in the original text.)

5 **In 2013...China:** ProMED-mail, "Undiagnosed viral pneumonia— China: (AH) medical staff, RFI," *ProMED-mail Archive 20130614.1773873*, June 14, 2013. Available at: www.promedmail.org.

**5 In 2006...China:** ProMED-mail, "Undiagnosed pneumonia—China (HK ex mainland): RFI," *ProMED-mail Archive 20060622.1734*, June 22, 2006. Available at: www.promedmail.org.

**6 In February 2003...(WHO):** Elisabeth Rosenthal with Lawrence K. Altman, "China raises tally of cases and deaths in mystery illness," *New York Times*, March 27, 2003, www.nytimes.com/2003/03/27/world /china-raises-tally-of-cases-and-deaths-in-mystery-illness.html.

**6 Chinese authorities...December 31st:** World Health Organization, "Pneumonia of unknown cause—China," January 5, 2020, www.who .int/csr/don/05-january-2020-pneumonia-of-unkown-cause-china/en.

**6 But by January...so far:** ProMED-mail, "Undiagnosed pneumonia— China (HU) (02): updates, other country responses, RFI," *ProMED-mail Archive 20200103.6869668*, January 3, 2020. Available at: www.promedmail.org.

**6 On January 8th...spread:** ProMED-mail, "Undiagnosed pneumonia—China (HU) (05): novel coronavirus identified," *ProMED-mail Archive 20200108.6877694*, January 8, 2020. Available at: www.promedmail.org.

**7 "critical public...wrong":** Jeremy Farrar, Twitter Post, January 10, 2020, 9:50 AM, twitter.com/JeremyFarrar/status/1215647022893670401.

**7 The Shanghai...spawned SARS:** Zhuang Pinghui, "Chinese laboratory that first shared coronavirus genome with world ordered to close for 'rectification,' hindering its Covid-19 research," *South China Morning Post*, February 28, 2020, www.scmp.com/news/china/society/article/3052966 /chinese-laboratory-first-shared-coronavirus-genome-world-ordered.

**7 On January 7th...the virus:** Andrew Rambaut, "Preliminary phylogenetic analysis of 11 nCoV2019 genomes, 2020-01-19," Virological, virological.org/t/preliminary-phylogenetic-analysis-of-11-ncov2019-genomes -2020-01-19/329.

**7 The Shanghai lab...next day:** Zhuang Pinghui, "Chinese laboratory that first shared coronavirus genome with world ordered to close for 'rectification,' hindering its Covid-19 research."

**9 In January, they...go to the hospital:** Natsuko Imai, et al., "Report 1—Estimating the potential total number of novel Coronavirus (2019-nCoV) cases in Wuhan City, China," MRC Centre for Global Infectious Disease Analysis, January 17, 2020, www.imperial.ac.uk/mrc-global-infectious -disease-analysis/covid-19/report-1-case-estimates-of-covid-19.

**9 On January 10th...came home:** Jasper Fuk-Woo Chan, et al., "A familial cluster of pneumonia associated with the 2019 novel coronavirus indicating person-to-person transmission: a study of a family cluster," *The Lancet* 395, no. 10223 (January 2020): 514–23, doi.org/10.1016/s0140-6736 (20)30154-9.

**9 On January 15th...sustained human-to-human transmission:** ProMED-mail, "Novel coronavirus (05): China (HU), Japan ex China," *ProMED-mail Archive 20200115.6891515*, January 15, 2020. Available at: www.promedmail.org.

**10 On January 18th...served:** *Sina*, news.sina.com.cn/s/2020-01-21 /doc-iihnzhha3843904.shtml.

**10 The mayor...limited:** James Kynge, Sun Yu, and Tom Hancock, "Coronavirus: the cost of China's public health cover-up," *Financial Times*, February 6, 2020, www.ft.com/content/fa83463a-4737-11ea-aeb3 -955839e06441.

**10 Then a local...correct here:** ProMED-mail, "Novel coronavirus (07): China (HU), Thailand ex China, Japan ex China, WHO," *ProMED-mail Archive 20200117.6895647*, January 17, 2020. Available at: www.promed mail.org.

**10 By January 20th...climbing:** ProMED-mail, "Novel coronavirus (11): China (HU), South Korea ex China," *ProMED-mail Archive 20200120.6899007*, January 20, 2020. Available at: www.promedmail.org.

**10 Also on...people:** *Caixin*, www.caixin.com/2020-01-20/101506222 .html.

**11 The *South China*...reported:** Josephine Ma, "Coronavirus: China's first confirmed Covid-19 case traced back to November 17," *South China Morning Post*, March 13, 2020, www.scmp.com/news/china/society/article /3074991/coronavirus-chinas-first-confirmed-covid-19-case-traced-back.

**11 The doctors...masks:** *Sina*, web.archive.org/web/20200411210210 /https://news.sina.com.cn/c/2020-02-08/doc-iimxyqvz1150881.shtml.

**14 Authorities later...enforced:** Josephine Ma and Zhuang Ping-hui, "5 million left Wuhan before lockdown, 1,000 new coronavirus cases expected in city," *South China Morning Post*, January 26, 2020, www.scmp .com/news/china/society/article/3047720/chinese-premier-li-keqiang -head-coronavirus-crisis-team-outbreak.

**14 Chris Dye...the 23rd:** Huaiyu Tian, et al., "An investigation of transmission control measures during the first 50 days of the COVID-19 epidemic in China," *Science*, March 31, 2020, doi.org/10.1126/science.abb6105.

**14 My first...global:** Debora MacKenzie, "New coronavirus looks set to cause a pandemic—how do we control it?" January 29, 2020, www.newscientist.com/article/2231864-new-coronavirus-looks-set-to-cause-a-pandemic-how-do-we-control-it.

**15 On January 27th...week:** MacKenzie, "New coronavirus looks set to cause a pandemic—how do we control it?"

**16 Three days...adapted to humans:** Chaolin Huang, et al., "Clinical features of patients infected with 2019 novel coronavirus in Wuhan, China," *Lancet* 395, no. 10223 (January 2020): 497–506, doi.org/10.1016/S0140-6736(20)30183-5.

**16 On March 11th...coronavirus:** Translation by Elisabeth Bik, "Dr. Ai Fen, 艾芬, the Wuhan Whistle," *Scientific Integrity Digest*, March 11, 2020, scienceintegritydigest.com/2020/03/11/dr-ai-fen-the-wuhan-whistle.

**17 Back in December...department:** Translation by Elisabeth Bik, "Dr. Ai Fen, 艾芬, the Wuhan Whistle."

**17 The news...censored:** Kynge, Yu, and Hancock, "Coronavirus: the cost of China's public health cover-up."

**17 The hospital told...alarm:** Translation by Elisabeth Bik, "Dr. Ai Fen, 艾芬, the Wuhan Whistle."

**17 The Japanese...rumors:** Keisuke Kawazu, "Public backlash over China gov't accusations against docs who sounded coronavirus alarm," *The Mainichi*, January 31, 2020, mainichi.jp/english/articles/20200131/p2a/00m/0in/021000c.

**18 And, that day...YY:** Lotus Ruan, Jeffrey Knockel, and Masashi Crete-Nishihata, "Censored contagion: how information on the coronavirus is managed on Chinese social media," *The Citizen Lab* (University of Toronto), March 3, 2020, citizenlab.ca/2020/03/censored-contagion-how-information-on-the-coronavirus-is-managed-on-chinese-social-media.

**18 If I had...the whistle:** Lily Kuo, "Coronavirus: Wuhan doctor speaks out against authorities," *Guardian*, March 11, 2020, www.theguardian.com/world/2020/mar/11/coronavirus-wuhan-doctor-ai-fen-speaks-out-against-authorities.

# Notes

**21 Andy Tatem...globally, they wrote:** Shengjie Lai, et al., "Effect of non-pharmaceutical interventions for containing the COVID-19 outbreak: an observational and modelling study," medRxiv preprint, March 9, 2020, doi.org/10.1101/2020.03.03.20029843.

**22 To stop...bureaucrats:** Steven Lee Myers, "China created a failsafe system to track contagions. It failed," *New York Times,* March 29, 2020, www.nytimes.com/2020/03/29/world/asia/coronavirus-china.html.

**25 more people...down stairs:** Phil Hammond, Twitter Post, January 24, 2020, 3:10 AM, twitter.com/drphilhammond/status/1220619993 408266241.

**26 Using...case infected:** Matt J Keeling, et al., "The efficacy of contact tracing for the containment of the 2019 novel coronavirus (COVID-19)," medRxiv preprint, February 17, 2020, doi.org/10.1101/2020 .02.14.20023036.

**26 Rosalind...symptoms:** Joel Hellewell, et al., "Feasibility of controlling COVID-19 outbreaks by isolation of cases and contacts," *The Lancet Global Health* 8 (February 2020): 488–96, doi.org/10.1016/S2214 -109X(20)30074-7.

**28 Zeng Guang...New Year:** Kynge, Yu, and Hancock, "Coronavirus: the cost of China's public health cover-up."

**32 In an astonishing...declared:** Lee Hsien Loong, "PM Lee Hsien Loong on the 2019-nCoV situation in Singapore," Facebook, February 8, 2020, www.facebook.com/watch/?v=1284271178628870.

**32 At the end...nearly half:** Benjamin J. Cowling, et al., "Impact Assessment of Non-Pharmaceutical Interventions against Coronavirus Disease 2019 and Influenza in Hong Kong: an Observational Study," *The Lancet Public Health* 5, no. 5 (April 2020), doi.org/10.1016/s2468-2667(20)30090-6.

**33 The small Italian...as needed:** Andrea Crisanti and Antonio Cassone, "In one Italian town, we showed mass testing could eradicate the coronavirus," *Guardian*, March 20, 2020, www.theguardian.com/com mentisfree/2020/mar/20/eradicated-coronavirus-mass-testing-covid -19-italy-vo.

**35 According to...contain the virus:** "Eight Wuhan residents praised for 'whistle-blowing' virus outbreak," *Global Times*, January 29, 2020, www .globaltimes.cn/content/1177960.shtml.

**36 In fact, Italian...January:** D. Cereda, et al., "The early phase of the COVID-19 outbreak in Lombardy, Italy," *arXiv* pre-print, March 20, 2020, arxiv.org/abs/2003.09320.

## CHAPTER 2

**39 In 1972...dull:** David S. Jones, "History in a crisis—lessons for Covid-19," *New England Journal of Medicine* 382, no. 18 (April 2020): 1681–1683, doi.org/10.1056/nejmp2004361.

**41 Investment...in 2019:** "AJPH editorial: US readiness for COVID-19, other outbreaks hinges on investments to public health system," American Public Health Association, February 13, 2020, www.apha.org/news-and-media/news-releases/ajph-news-releases/2020/ajph-editorial.

**41 There has been a surge...at that time:** Melinda Wenner Moyer, "A Wave of Resurgent Epidemics Has Hit the U.S.," *Scientific American,* May 1, 2018, www.scientificamerican.com/article/a-wave-of-resurgent-epidemics-has-hit-the-u-s.

**42 In Europe...Covid-19:** Chris Thomas, "Hitting the poorest worst? How public health cuts have been experienced in England's most deprived communities," Institute for Public Policy Research, May 11, 2019, www.ippr.org/blog/public-health-cuts#anounce-of-prevention-is-worth-a-pound-of-cure.

**42 According to virologist...human deaths:** Ab Osterhaus and Leslie Reperant, "Emerging and re-Emerging Viruses: Origins and Drivers," European Society for Virology, April 11, 2016, www.eusv.eu/emerging-and-re-emerging-viruses-origins-and-drivers.

**43 In the 1800s...novels:** "Contagion: Historical Views of Diseases and Epidemics," Harvard Library, ocp.hul.harvard.edu/contagion/tuberculosis.html.

**43 Yellow fever...New World:** "Contagion: historical views of diseases and epidemics," Harvard Library, ocp.hul.harvard.edu/contagion/tuberculosis.html.

**43 By 2004...few percent:** Rafael Lozano, et al., "Global and regional mortality from 235 causes of death for 20 age groups in 1990 and 2010: a systematic analysis for the Global Burden of Disease Study 2010," *Lan-*

*cet* 380 (2012): 2095-128, ipa-world.org/society-resources/code/images /95b1494-Lozano%20Mortality%20GBD2010.pdf.

**46 HIV had probably…most infections were there:** Nuno R. Faria, et al., "The early spread and epidemic ignition of HIV-1 in human populations," *Science* 346, no. 6205 (October 2014): 56–61, doi.org/10.1126 /science.1256739.

**47 Jacques Pépin…with HIV:** Jacques Pépin, *The Origin of AIDS* (Cambridge: Cambridge UP, 2011).

**48 in 1992…cost-effective:** Institute of Medicine (US) Committee on Emerging Microbial Threats to Health, *Emerging Infections: Microbial Threats to Health in the United States*, eds. Joshua Lederberg, Robert E. Shope, and Stanley C. Oaks, Jr. (Washington, DC: National Academies Press, 1992), doi.org/10.17226/2008.

**48 In 2016…per year:** Commission on a Global Health Risk Framework for the Future, National Academy of Medicine, Secretariat, *The Neglected Dimension of Global Security: A Framework to Counter Infectious Disease Crises* (Washington, DC: National Academies Press, 2016), doi .org/10.17226/21891.

**48 The rinderpest…pigs:** Yuki Furuse, et al., "Origin of measles virus: divergence from rinderpest virus between the 11th and 12th centuries," *Virology Journal* 7, no. 1 (March 2010): 52, doi.org/10.1186/1743 -422x-7-52.

**49 mumps…pigs:** Nathan D. Wolfe, et al., "Origins of Major Human Infectious Diseases." *Nature* 447, no. 7142 (May 2007): 279–83. doi.org/10.1038 /nature05775.

**50 Then, in 1997…reds do not:** Debora MacKenzie, "Sick to death," *New Scientist,* August 5, 2020, www.newscientist.com/article/mg16722504 -300-sick-to-death.

**50 In 2002…ill effects:** Debora MacKenzie, "Plague on a national icon," *New Scientist,* October 26, 2002, www.newscientist.com/article /mg17623661-100-plague-on-a-national-icon.

**50 In 1998…extinct:** L. Berger, et al., "Chytridiomycosis causes amphibian mortality associated with population declines in the rain forests of Australia and Central America," *Proceedings of the National*

*Academy of Sciences* 95, no. 15 (July 1998): 9031–36, doi.org/10.1073 /pnas.95.15.9031.

**50 In 2008…wildlife:** Kate E. Jones, et al., "Global trends in emerging infectious diseases," *Nature* 451, no. 7181 (2008): 990–93, doi.org/10.1038 /nature06536.

**52 Fabian Leendertz…they died:** Almudena Marí Saéz, et al., "Investigating the zoonotic origin of the West African Ebola epidemic," *EMBO Mol Med* 7, no. 1 (January 2015), doi.org/10.15252/emmm.201404792.

**54 The one exception…writers:** Paul Nuki and Alanna Shaik, "Scientists put on alert for deadly new pathogen—'Disease X'," *Telegraph*, March 10, 2018, www.telegraph.co.uk/global-health/science-and-disease/world -health-organization-issues-alert-disease-x.

**55 according…aggressive:** "Factsheet about Crimean-Congo haemorrhagic fever," European Centre for Disease Prevention and Control (EU), www.ecdc.europa.eu/en/crimean-congo-haemorrhagic-fever/facts /factsheet.

**55 Meanwhile, the virus…in Spain:** Ana Negredo, et al., "Survey of Crimean-Congo hemorrhagic fever enzootic focus, Spain, 2011–2015," *Emerging Infectious Diseases* 25, no. 6 (June 2019): 1177–84, doi.org/10.3201 /eid2506.180877.

**56 In 2008…Zambia:** Debora MacKenzie, "New killer virus makes an appearance," *New Scientist*, October 15, 2008, www.newscientist.com /article/mg20026783-200-new-killer-virus-makes-an-appearance.

**57 Just as they had…infected person:** Nuno Rodrigues Faria, et al., "Zika virus in the Americas: early epidemiological and genetic findings," *Science* 352, no. 6283 (April 2016): 345–49, doi.org/10.1126/science.aaf5036.

**60 In 1998…it was bats:** Lai-Meng Looi, "Lessons from the Nipah virus outbreak in Malaysia," *Malaysian Journal of Pathology* 29, no. 2 (2007): 63–67, www.mjpath.org.my/2007.2/02Nipah_Virus_lessons.pdf.

**62 They are being…Covid-19:** Chunyan Wang, et al., "A Human Monoclonal Antibody Blocking SARS-CoV-2 Infection," *Nature Commnications* 11, no. 2251, May 12, 2020, doi.org/10.1101/2020.03.11.987958.

**62 In 2014, Daszak…infected:** Olivier Pernet, et al., "Evidence for Henipavirus Spillover into Human Populations in Africa," *Nature Communications* 5, no. 1 (November 2014), doi.org/10.1038/ncomms6342.

**62 It started...heels:** Debora MacKenzie, "World must get ready now for the next big health threat," *New Scientist*, December 15, 2015, www .newscientist.com/article/mg22830522-900-world-must-get-ready-now -for-the-next-big-health-threat.

**64 Yet it should...transmitting:** Debora MacKenzie, "Ebola rapidly evolves to be more transmissible and deadlier," *New Scientist*, November 3, 2016, www.newscientist.com/article/2111311-ebola-rapidly-evolves-to -be-more-transmissible-and-deadlier.

## CHAPTER 3

**66 Franklin Jones...SARS:** Nanshan Zhong and Guangqiao Zeng, "What we have learnt from SARS epidemics in China," *BMJ* 333, no. 7564 (August 2006): 389–91, doi.org/10.1136/bmj.333.7564.389.

**66 The same day...both messages:** ProMED-mail, "Pneumonia-China (Guangdong): RFI," *ProMED-mail Archive 20030210.0357*, February 10, 2003. Available at: www.promedmail.org.

**68 The same day...nurses:** ProMED-mail, "Pneumonia—China (Guangdong) (03)," *ProMED-mail Archive 20030214.039*, February 14, 2003. Available at: www.promedmail.org.

**68 On February 18th...dubious:** ProMED-mail, "Pneumonia—China (Guangdong) (04)," *ProMED-mail Archive 20030219.0427*, February 19, 2003. Available at: www.promedmail.org.

**68 On the 20th...out:** ProMED-mail, "Pneumonia—China (Guangdong) (06)," *ProMED-mail Archive 20030220.0447*, February 20, 2003. Available at: www.promedmail.org.

**70 Both viruses...inflammation:** Meredith Wadman, Jennifer Couzin-Frankel, Jocelyn Kaiser, and Catherine Matacic, "How does coronavirus kill? Clinicians trace a ferocious rampage through the body, from brain to toes," *Science*, April 17, 2020, www.sciencemag.org/news/2020/04 /how-does-coronavirus-kill-clinicians-trace-ferocious-rampage-through -body-brain-toes.

**71 But as fears...to disease:** Christian Kreuder-Sonnen, "China vs the WHO: a Behavioural Norm Conflict in the SARS Crisis," International Affairs 95, no. 3 (January 2019): 535–52, doi.org/10.1093/ia/iiz022.

**72 The next day...controlled:** Tim Brookes with Omar A. Khan,

*Behind the Mask: How the World Survived SARS, the First Epidemic of the 21st Century* (Washington, DC: American Public Health Association, 2005), 195.

**73 Yanzhong Huang...37:** Yanzhong Huang, "The SARS Epidemic and its Aftermath in China: A Political Perspective," Learning from SARS: Preparing for the Next Disease Outbreak: Workshop Summary, Eds. Stacey Knobler, et al (Washington, DC: National Academes Press, 2004), www.ncbi.nlm.nih.gov/books/NBK92479.

**74 Huang wrote...February 11th:** Yanzhong Huang, "The SARS Epidemic and its Aftermath in China: A Political Perspective."

**77 In its world...epidemic:** World Health Organization, *The World Health Report 2003: Shaping the Future* (Geneva, Switzerland: WHO, 2003), www.who.int/whr/2003/en.

**83 In 2005...as a threat:** Mark Henderson, "End of Sars as a deadly threat," *Times of London*, February 21, 2009, www.thetimes.co.uk/article /end-of-sars-as-a-deadly-threat-nz3ll7tqzsz.

**84 But, said virologists...humans were:** L. F. Wang and B. T. Eaton, "Bats, Civets and the Emergence of SARS," *Current Topics in Microbiology and Immunology Wildlife and Emerging Zoonotic Diseases: The Biology, Circumstances and Consequences of Cross-Species Transmission*, (2007): 325–44, doi.org/10.1007/978-3-540-70962-6_13.

**84 Also, that year...markets:** Zhang Feng, "Does SARS virus still exist in the wild?" *China Daily*, February 23, 2005, www.chinadaily.com.cn /english/doc/2005-02/23/content_418481.htm.

**84 If no action...strain:** Nanshan Zhong and Guangqiao Zeng, "What we have learnt from SARS epidemics in China."

**85 he discovered...ProMED:** ProMED-mail, "Novel coronavirus—Saudi Arabia: human isolate," *ProMED-mail Archive 20120920.1302733*, September 20, 2012. Available at: www.promedmail.org.

**85 Within days...Jeddah:** Debora MacKenzie, "Threatwatch: Find the germs, don't sack the messenger," New Scientist, October 24, 2012, www .newscientist.com/article/dn22417-threatwatch-find-the-germs-dont -sack-the-messenger.

**86 It was in local...camels:** Kate Kelland, "Special Report—Saudi Arabia takes heat for spread of MERS virus," Reuters, May 22, 2014,

uk.reuters.com/article/uk-saudi-mers-special-report/special-report-saudi
-arabia-takes-heat-for-spread-of-mers-virus-idUKKBN0E207Z20140522.

**87 Last year, epidemiologists...since 2016:** Christl A. Donnelly, et al.,
"Worldwide Reduction in MERS Cases and Deaths since 2016," *Emerging
Infectious Diseases* 25, no. 9 (September 2019): 1758–60, doi.org/10.3201
/eid2509.190143.

**88 When I reported...denial:** Debora MacKenzie, "Secrets and Lies
in Europe," *New Scientist,* May 3, 1997, www.newscientist.com/article
/mg15420802-300-secrets-and-lies-in-europe.

## CHAPTER 4

**90 So does...worldwide:** World Health Organization, "Global Hepatitis Report, 2017," 2017, apps.who.int/iris/handle/10665/255016.

**90 In April...Myanmar:** Marc T. Valitutto, et al., "Detection of novel
coronaviruses in bats in Myanmar," *PLoS One* 15, no. 4 (April 2020):
e0230802, doi.org/10.1371/journal.pone.0230802.

**90 In 2017...evolution:** Simon J. Anthony, et al., "Global patterns
in coronavirus diversity," *Virus Evolution* 3, no. 1 (January 2017), doi.org
/10.1093/ve/vex012.

**91 Wildlife scientists...discovery:** Anthony King, "Super bats: What
doesn't kill them, could make us stronger," *New Scientist,* February 10, 2016,
www.newscientist.com/article/2076598-super-bats-what-doesnt-kill
-them-could-make-us-stronger.

**91 But a 2017...mammals:** Kevin J. Olival, et al., "Host and Viral
Traits Predict Zoonotic Spillover from Mammals," *Nature* 546, no. 7660
(June 2017): 646–50, doi.org/10.1038/nature22975.

**93 Sure enough...Covid-19 uses:** Wendong Li, et al., "Bats are natural reservoirs of SARS-like coronaviruses," *Science* 310, no. 5748 (October
2005): 676–79, doi.org/10.1126/science.1118391.

**95 In 2013...team concluded:** Xing-Yi Ge, et al., "Isolation and
characterization of a bat SARS-like coronavirus that uses the ACE2
receptor," *Nature* 503, no. 7477 (October 2013): 535–38, doi.org/10.1038
/nature12711.

**95 In 2017...they warned:** Ben Hu, et al., "Discovery of a rich gene
pool of bat SARS-related coronaviruses provides new insights into the

origin of SARS coronavirus," *PLoS Pathogens* 13, no. 11 (November 2017), doi.org/10.1371/journal.ppat.1006698.

95 **The title...SARS-like viruses:** Vineet D. Menachery, et al., "A SARS-like Cluster of Circulating Bat Coronaviruses Shows Potential for Human Emergence," *Nature Medicine* 21, no. 12 (November 2015): 1508–13, doi.org/10.1038/nm.3985.

96 **In 2016...vaccines:** Vineet D. Menachery, et al., "SARS-like WIV1-CoV poised for human emergence," *Proceedings of the National Academy of Sciences* 113, no. 11 (March 2016): 3048–53, doi.org/10.1073/pnas.1517719113.

96 Debora MacKenzie, "Plague! How to prepare for the next pandemic," *New Scientist,* February 22, 2017, www.newscientist.com/article/mg23331140-400-plague-how-to-prepare-for-the-next-pandemic/#ixzz6KMAMFWDf.

96 **In 2018, Shi's...happened:** Ning Wang, et al., "Serological Evidence of Bat SARS-Related Coronavirus Infection in Humans, China." *Virologica Sinica* 33, no. 1 (February 2018): 104–7, doi.org/10.1007/s12250-018-0012-7.

96 **It is highly...signs:** Yi Fan, et al., "Bat coronaviruses in China," *Viruses* 11, no. 3 (March 2019): 210, doi.org/10.3390/v11030210.

97 **Perhaps the team's...implemented:** Peng Zhou, et al., "A pneumonia outbreak associated with a new coronavirus of probable bat origin," *Nature* 579, no. 7798 (February 2020): 270–73, doi.org/10.1038/s41586-020-2012-7.

100 **Related viruses...people:** Tommy Tsan-Yuk Lam, et al., "Identifying SARS-CoV-2 Related Coronaviruses in Malayan Pangolins," *Nature,* March 26, 2020, doi.org/10.1038/s41586-020-2169-0.

101 **By late March...re-opening:** Bloomberg News, "Wuhan is returning to life. So are its disputed wet markets," *Bloomberg,* April 8, 2020, www.bloomberg.com/news/articles/2020-04-08/wuhan-is-returning-to-life-so-are-its-disputed-wet-markets.

102 **Bats are traditionally...rarer:** Tammy Mildenstein, Iroro Tanshi, and Paul A. Racey, "Exploitation of bats for bushmeat and medicine," in *Bats in the Anthropocene: Conservation of Bats in a Changing World,* eds. Christian C. Voigt and Tigga Kingston (Cham, Switzerland: Springer Open, 2016), doi.org/10.1007/978-3-319-25220-9_12.

102 **One, charging...into use:** "Ye Ming Sha, bat feces, bat dung, bat

guano," Best Plant, www.bestplant.shop/products/ye-ming-sha-bat-feces
-bat-dung-bat-guano.

102 The **Clinical**...vitamin A: Chun-Han Zhu, *Clinical Handbook of Chinese Prepared Medicines* (Brookline, MA: Paradigm, 1989), 179.

103 An online site...at night: Peter Borten, "Chinese herbs," chi neseherbinfo.com/ye-ming-sha-bat-feces.

103 Sampling...persistent there: Francesca Colavita, et al., "SARS-CoV-2 isolation from ocular secretions of a patient with COVID-19 in Italy with prolonged viral RNA detection," *Annals of Internal Medicine* [Epub ahead of print 17 April 2020], doi.org/10.7326/M20-1176.

103 and that eyes...infection: Kenrie P. Y. Hui, "Tropism, replication competence, and innate immune responses of the coronavirus SARS-CoV-2 in human respiratory tract and conjunctiva: an analysis in ex-vivo and in-vitro cultures," *The Lancet Respiratory Medicine*, May 7, 2020, doi.org/10.1016/S2213-2600(20)30193-4.

104 than were used...burned alive: Newsflare, "Hundreds of bats burned in Indonesia in bid to prevent coronavirus spread," Yahoo! News, March 16, 2020, news.yahoo.com/hundreds-bats-burned-indonesia-bid-150000233.html.

104 The use of...cures: Jani Actman, "Traditional Chinese medicine and wildlife," National Geographic, February 7, 2019, www.nationalgeo graphic.com/animals/reference/traditional-chinese-medicine.

106 All these chemical...40: Duke-NUS Graduate Medical School, "Researchers Find Genetic Link Between Bats' Ability to Fly and Viral Immunity," Duke Global Health Institute, December 20, 2012, globalhealth .duke.edu/news/researchers-find-genetic-link-between-bats-ability-fly -and-viral-immunity.

106 Bats' high...infection: Jiazheng Xie, et al., "Dampened STING-Dependent Interferon Activation in Bats," *Cell Host & Microbe* 23, no. 3 (March 2018), doi.org/10.1016/j.chom.2018.01.006.

107 in February 2020...cells: Cara E. Brooke, et al., "Accelerated viral dynamics in bat cell lines, with implications for zoonotic emergence," *eLife* (February 2020), doi.org/10.7554/eLife.48401.

109 Simply...international: "Bat Conservation International bats and disease position statement," Bats & Human Health, Bat Conservation

International, www.batcon.org/resources/for-specific-issues/bats-human
-health.

**109 in 2006...by bats:** Charles H. Calisher, et al., "Bats: important reservoir hosts of emerging viruses," *Clinical Microbiology Reviews* 19, no. 3
(July 2006): 531–45, doi.org/10.1128/cmr.00017-06.

**110 There has been...in 2015:** Raina K. Plowright, et al., "Ecological dynamics of emerging bat virus spillover," *Proceedings of the Royal
Society B: Biological Sciences* 282, no. 1798 (January 7, 2015): 20142124, doi
.org/10.1098/rspb.2014.2124.

**110 In 2008...more of it:** Raina K. Plowright, et al., "Reproduction
and Nutritional Stress Are Risk Factors for Hendra Virus Infection in Little
Red Flying Foxes (Pteropus Scapulatus)," *Proceedings of the Royal Society
B: Biological Sciences* 275, no. 1636 (January 2008): 861–69, doi.org/10.1098
/rspb.2007.1260.

## CHAPTER 5

**114 His lab...disease:** Ron A.M. Fouchier, et al., "Koch's postulates ful-
filled for SARS virus," *Nature* 423 (May 2003): 240, doi.org/10.1038/423240a.

**116 Derek Smith...Hemisphere's winter:** Colin A. Russell, et al.,
"The global circulation of seasonal influenza A (H3N2) viruses," *Science*
320, no. 5874 (April 2008), doi.org/10.1126/science.1154137.

**118 In Australia...protection:** Debora MacKenzie, "Jab in the dark:
Why we don't have a universal flu vaccine," *New Scientist*, January 2, 2018,
www.newscientist.com/article/2156915-jab-in-the-dark-why-we-dont
-have-a-universal-flu-vaccine.

**121 In 2004, virologist...used to:** R.J. Webby, et al., "Multiple lineages
of antigenically and genetically diverse influenza A virus co-circulate in
the United States swine population," *Virus Research* 103, no. 1–2 (July
2004): 67–73, doi.org/10.1016/j.virusres.2004.02.015.

**122 Five...imminent:** Laura MacInnis and Stephanie Nebehay,
"WHO warns flu pandemic imminent," *Reuters*, April 28, 2009, www.reu
ters.com/article/us-flu/who-warns-flu-pandemic-imminent-idUSTRE
53N22820090429.

**123 The UN...to pigs:** "FAO acts over H1N1 human crisis," Food and

Agriculture Organization of the United Nations, April 27, 2009, www.fao
.org/news/story/en/item/13002/icode.

**124 On June 11th...pandemic:** "WHO pandemic declaration," Centers for Disease Control and Prevention, www.cdc.gov/h1n1flu/who.

**127 I think...CDC:** Richard Knox, "Flu pandemic much milder than expected," *NPR Morning Edition*, December 8, 2009, www.npr.org/tem
plates/story/story.php?storyId=121184706.

**127 in March...Covid-19:** "COVID-19 pandemic just started, hard to see end: Chinese epidemiologist," *Global Times*, March 24, 2020, www
.globaltimes.cn/content/1183619.shtml.

**129 When it hit...another one:** Public Health England, "Pandemic Influenza Response Plan 2014," August 2014, assets.publishing.service
.gov.uk/government/uploads/system/uploads/attachment_data/file
/344695/PI_Response_Plan_13_Aug.pdf.

**130 In 1999...Influenza Viruses:** Angela N. Cauthen, et al., "Continued circulation in China of highly pathogenic avian influenza viruses encoding the hemagglutinin gene associated with the 1997 H5N1 outbreak in poultry and humans," *Journal of Virology* 74, no. 14 (July 2000): 6592–99, doi.org/10.1128/jvi.74.14.6592-6599.2000.

**130 In 2002...pandemic concern:** Y. Guan, et al., "Emergence of multiple genotypes of H5N1 avian influenza viruses in Hong Kong SAR," *Proceedings of the National Academy of Sciences* 99, no. 13 (June 2002): 8950–55, doi.org/10.1073/pnas.132268999.

**131 On the 28th...unseen:** Debora Mackenzie, "Bird flu outbreak started a year ago," *New Scientist*, January 28, 2004, www.newscientist
.com/article/dn4614-bird-flu-outbreak-started-a-year-ago.

**131 It is purely...surveillance:** Reuters, "China denies bird flu cover-up," *CNN International*, January 29, 2004, edition.cnn.com/2004
/WORLD/asiapcf/01/28/bird.flu.china.reut.

**132 Sure enough...report the deaths:** Oliver August, "China covers up again on outbreak," *The Times*, February 2, 2004, www.thetimes.co.uk
/article/china-covers-up-again-on-outbreak-hntz3rp3rgj.

**133 By 2006, Yi...trade:** H. Chen, et al., "Establishment of multiple sublineages of H5N1 influenza virus in Asia: implications for pandemic control," *Proceedings of the National Academy of Sciences* 103, no. 8 (February 2006): 2845–50, doi.org/10.1073/pnas.0511120103.

**134 Yi Guan found…southeastern China:** H. Chen, et al., "H5N1 virus outbreak in migratory waterfowl," *Nature* 436, no. 7048 (July 2005): 191 –92, doi.org/10.1038/nature03974.

**135 Senior Chinese…samples:** Debora MacKenzie, "China denies bird flu research findings," *New Scientist*, July 13, 2005, www.newscientist .com/article/mg18725083-500-china-denies-bird-flu-research-findings.

**135 There was…February 2020:** Cissy Zhou, "China reports outbreak of deadly bird flu among chickens in Hunan province, close to coronavirus epicentre of Wuhan," *South China Morning Post*, February 2, 2020, www.scmp.com/news/china/society/article/3048566/china-reports -outbreak-deadly-bird-flu-among-chickens-hunan.

**136 In Southeast Asia…by 2005:** Anni McLeod, et al., "Economic and social impacts of avian influenza," FAO Emergency Centre for Transboundary Animal Diseases Operations (ECTAD), November 2005, www.fao.org /avianflu/documents/Economic-and-social-impacts-of-avian-influenza -Geneva.pdf.

**136 Since 2013…October 2017:** Public Health England, "Risk assessment of avian influenza A(H7N9)—eighth update," January 8, 2020, www .gov.uk/government/publications/avian-influenza-a-h7n9-public-health -england-risk-assessment/risk-assessment-of-avian-influenza-ah7n9 -sixth-update.

**137 So Ron Fouchier…its deadliness:** S. Herfst, et al., "Airborne transmission of influenza A/H5N1 virus between ferrets," *Science* 336, no. 6088 (June 21, 2012): 1534–41, doi.org/10.1126/science.1213362.

**138 In 2017…does that:** Masaki Imai, et al., "A Highly Pathogenic Avian H7N9 Influenza Virus Isolated from A Human Is Lethal in Some Ferrets Infected via Respiratory Droplets," *Cell Host & Microbe* 22, no. 5 (November 2017), doi.org/10.1016/j.chom.2017.09.008.

**139 In 2012…fund them:** Anthony S. Fauci, "Research on highly pathogenic H5N1 influenza virus: the way forward," *MBio*3, no. 5 (October 2012), doi.org/10.1128/mbio.00359-12.

**139 In 2017…to resume:** National Institutes of Health, "Notice announcing the removal of the funding pause for gain-of-function research projects," December 19, 2017, grants.nih.gov/grants/guide/notice-files/NOT-OD-17-071.html.

**139 In 2019, NIAID…human cells:** Peter Daszak (EcoHealth Alliance), "Understanding the risk of bat coronavirus emergence," Project Number: 2R01AI110964-06, NIH Research Portfolio Online Reporting Tools (RePORT), projectreporter.nih.gov/project_info_description.cfm?aid=9819304&icde=49645421.

**140 EcoHealth…in the USA:** EcoHealth Alliance, "Regarding NIH termination of coronavirus research funding," April 2020, www.ecohealthalliance.org/2020/04/regarding-nih-termination-of-coronavirus-research-funding.

## CHAPTER 6

**141 The world…for war:** Bill Gates, "Innovation for pandemics," *The New England Journal of Medicine* 378 (May 2018): 2057–60, doi.org/0.1056/NEJMp1806283. Remarks originally delivered as the Shattuck Lecture for the Massachusetts Medical Society on April 27, 2018.

**142 Christopher Kirchhoff…collapse:** Christopher Kirchhoff, "Memorandum for Ambassador Susan E. Rice, Subject: NSC Lessons Learned Study on Ebola," National Security Council, White House, July 11, 2016, assets.documentcloud.org/documents/6817684/NSC-Ebola-Lessons-Learend-Report-FINAL-8-28-16.pdf.

**142 The US tried…Trump administration:** Christopher Kirchhoff, "Ebola should have immunized the United States to the coronavirus," *Foreign Affairs*, March 28, 2020, www.foreignaffairs.com/articles/2020-03-28/ebola-should-have-immunized-united-states-coronavirus.

**142 On March 11th…inaction:** Tedros Ghebreyesus, "WHO director-general's opening remarks at the media briefing on COVID-19," March 11, 2020, www.who.int/dg/speeches/detail/who-director-general-s-opening-remarks-at-the-media-briefing-on-covid-19---11-march-2020.

**143 senior officials…were right:** Yasmeen Abutaleb, Josh Dawsey, Ellen Nakashima, and Greg Miller, "The U.S. was beset by denial and dysfunction as the coronavirus raged," *Washington Post*, April 4, 2020, www.washingtonpost.com/national-security/2020/04/04/coronavirus-government-dysfunction.

# Notes

143 **A Global...to act:** Global Preparedness Monitoring Board, "A world at risk: annual report on global preparedness for health emergencies," September 2019, apps.who.int/gpmb/assets/annual_report/GPMB_Annual_Report_English.pdf.

144 **A high-level...insufficient:** United Nations, High-Level Panel on the Global Response to Health, "Protecting humanity from future health crises: report of the High-Level Panel on the Global Response to Health Crises," February 2016, www.un.org/ga/search/view_doc.asp?symbol=A/70/723.

144 **After Ebola...48 hours:** "UK forms global infection response team," *BBC News*, November 1, 2016, www.bbc.com/news/health-37827388.

145 **The GPMB...gowns:** Global Preparedness Monitoring Board, "A world at risk: annual report on global preparedness for health emergencies."

146 **On March 26th...diagnostics and treatments:** G20, "G20 leaders' statement, extraordinary G20 leaders' summit statement on COVID-19," March 26, 2020, g20.org/en/media/Documents/G20_Extraordinary%20G20%20Leaders%E2%80%99%20Summit_Statement_EN%20(3).pdf.

146 **In 2018...RNA viruses:** The Johns Hopkins Center for Health Security, "The characteristics of pandemic pathogens," 2018, www.centerforhealthsecurity.org/our-work/pubs_archive/pubs-pdfs/2018/180510-pandemic-pathogens-report.pdf.

147 **Fortunately, there...track:** Debora MacKenzie, "Germ detectors: Unmasking our microbial foes," *New Scientist*, August 17, 2011, www.newscientist.com/article/mg21128262-400-germ-detectors-unmasking-our-microbial-foes.

152 **No amount...in 2018:** Edward C. Holmes, Andrew Rambaut, and Kristian G. Andersen, "Pandemics: spend on surveillance, not prediction," *Nature* 558, no. 7709 (June 7, 2018): 180–82, doi.org/10.1038/d41586-018-05373-w.

152 **monitor...populations:** "Our Approach," Global Virome Project, www.globalviromeproject.org/our-approach.

153 **The 2005 version...dangerous:** World Health Organization, "International Health Regulations, 2nd edition," 2005, www.who.int/ihr/9789241596664/en/www.who.int/ihr/9789241596664/en.

154 **The WHO was...fix:** Sarah Boseley, "World Health Organisation 'intentionally delayed declaring Ebola emergency,'" *Guardian*, March

20, 2015, www.theguardian.com/world/2015/mar/20/ebola-emergency
-guinea-epidemic-who.

**155 An assessment…52 percent:** Global Health Security Index, "2019
GHS Index," 2019, www.ghsindex.org/wp-content/uploads/2019/10/2019
-Global-Health-Security-Index.pdf.

**156 A similar flu…with Covid-19:** David E. Sanger, Eric Lipton, Eileen
Sullivan and Michael Crowley, "Before Virus Outbreak, a Cascade of Warnings Went Unheeded," *New York Times*, March 22, 2020, www.nytimes
.com/2020/03/19/us/politics/trump-coronavirus-outbreak.html.

**156 But it was…fast enough:** Lawrence O. Gostin, and Eric A. Fried-
man, "Ebola: a Crisis in Global Health Leadership," *The Lancet* 384, no.
9951 (October 2014): 1323–25, doi.org/10.1016/s0140-6736(14)61791-8.

**157 In March…small undertaking:** Scott Gottlieb et al., "National
coronavirus response: A road map to reopening," American Enterprise
Institute, March 29, 2020, www.aei.org/research-products/report/national
-coronavirus-response-a-road-map-to-reopening.

**159 There was a collective…experiments happened:** Debora Mac-
Kenzie, "US develops lethal new viruses," *New Scientist*, October 29, 2003,
www.newscientist.com/article/dn4318-us-develops-lethal-new-viruses.

**160 In February…virus like this:** Kristian G. Andersen, et al., "The
proximal origin of SARS-CoV-2," *Nature Medicine* 26, no. 4 (March 17,
2020): 450–52, doi.org/10.1038/s41591-020-0820-9.

**160 In March…frontline!:** Charles Calisher, et al., "Statement in support of the scientists, public health professionals, and medical professionals of China combatting COVID-19." *The Lancet* 395, no. 10226 (February
2020), doi.org/10.1016/s0140-6736(20)30418-9.

**162 In a paper…the public:** Albert D.M.E. Osterhaus, et al., "Make
science evolve into a One Health approach to improve health and
security: a white paper," *One Health Outlook* 2, no. 6 (2020), doi.org
/10.1186/s42522-019-0009-7.

**163 the world can…pandemic vaccine:** Kenneth A. Mclean, et al.,
"The 2015 global production capacity of seasonal and pandemic influenza
vaccine," *Vaccine* 34, no. 45 (October 2016): 5410–13, doi.org/10.1016/j
.vaccine.2016.08.019.

**169 Bill Gates…discarded:** Isobel Asher Hamilton, "Bill Gates is helping fund new factories for 7 potential coronavirus vaccines, even though it will waste billions of dollars," *Business Insider*, April 3, 2020, www .businessinsider.com/bill-gates-factories-7-different-vaccines-to-fight -coronavirus-2020-4.

**169 In late March...emergency:** Scott Gottlieb et al., "National coronavirus response: a road map to reopening."

**170 CEPI agrees...for that:** Coalition for Epidemic Preparedness Innovations, "Landmark global collaboration launched to defeat COVID-19 pandemic," April 24, 2020, cepi.net/news_cepi/landmark-global-collaboration -launched-to-defeat-covid-19-pandemic.

**171 One critic...stiff whiskey:** Debora MacKenzie, "Evidence that Tamiflu reduces deaths in pandemic flu," *New Scientist*, June 24, 2013, www.newscientist.com/article/dn23744-evidence-that-tamiflu-reduces -deaths-in-pandemic-flu.

**172 Jonathan Van...effect:** S.G. Muthuri, et al., "Impact of neuraminidase inhibitor treatment on outcomes of public health importance during the 2009-2010 influenza A (H1N1) pandemic: a systematic review and meta-analysis in hospitalized patients," *The Journal of Infectious Diseases* 207, no. 4 (November 2012): 553–63, doi.org/10.1093/infdis/jis726.

**175 In an investigation...infect bacteria:** Debora MacKenzie, "The war against antibiotic resistance is finally turning in our favour," *New Scientist*, January 16, 2019, www.newscientist.com/article/2190957-the-war -against-antibiotic-resistance-is-finally-turning-in-our-favour.

**176 In 2014...global GDP:** The Review on Antimicrobial Resistance (chaired by Jim O'Neill), "Antimicrobial resistance: tackling a crisis for the health and wealth of nations," December 2014, amr-review.org/sites /default/files/AMR%20Review%20Paper%20-%20Tackling%20a%20 crisis%20for%20the%20health%20and%20wealth%20of%20nations_1 .pdf.

**179 In 2006, California...ventilators:** Carla Marinucci, "Schwarzenegger: 'Shortsighted' for California to defund pandemic stockpile he built," *Politico*, March 31, 2020, www.politico.com/states/california/story/2020/03/31 /schwarzenegger-shortsighted-for-california-to-defund-pandemic -stockpile-he-built-1269954.

# Notes

180 **Guaranteeing that right…standard:** International Labour Organization (UN), COVID-19 and the world of work," www.ilo.org/global /topics/coronavirus/impacts-and-responses/WCMS_739049/lang--en /index.htm

180 **The GPMB…$53 billion:** Caroline Huber, et al., "The economic and social burden of the 2014 Ebola outbreak in West Africa," *The Journal of Infectious Diseases* 218, supplement 5 (October 2018), doi.org/10.1093 /infdis/jiy213.

180 **cancer deaths…hospitals:** Denis Campbell and Caroline Bannock, "Coronavirus crisis could lead to 18,000 more cancer deaths, experts warn," *Guardian*, April 28, 2020, www.theguardian.com/society/2020 /apr/29/extra-18000-cancer-patients-in-england-could-die-in-next-year -study.

180 **Epidemiologists at Imperial…respectively:** Alexandra B. Hogan, et al., "Report 19 - The Potential Impact of the COVID-19 Epidemic on HIV, TB and Malaria in Low- and Middle-Income Countries," Imperial College London, May 1, 2020, www.imperial.ac.uk/mrc-global-infectious-disease-analysis /covid-19/report-19-hiv-tb-malaria.

180 **A repeat…recession:** Olga B. Jonas (The World Bank), "Background paper: pandemic risk," *World Development Report*, October 2013, www.worldbank.org/content/dam/Worldbank/document/HDN/Health /WDR14_bp_Pandemic_Risk_Jonas.pdf.

180 **Now some economists…depression:** Nouriel Roubini, "The coming greater depression of the 2020s," *Project Syndicate*, April 28, 2020, www.project-syndicate.org/commentary/greater-depression-covid19 -headwinds-by-nouriel-roubini-2020-04.

181 **Three years ago…seriously:** Debora MacKenzie, "Plague! How to prepare for the next pandemic."

181 **$49 billion a year:** Congressional Budget Office, "Projected costs of U.S. nuclear forces, 2019 to 2028," January 24, 2019, www.cbo.gov /publication/54914.

181 **Yet this year…on the WHO:** World Health Organization, "Programme budget 2020-2021," 2019, www.who.int/about/finances-accoun tability/budget/en.

# Notes

## CHAPTER 7

**182 So, it turns out...Who knew?:** "The Most Important Jobs T-Shirt," Red Molotov, www.redmolotov.com/important-jobs-tshirt.

**183 A few years...complexity:** Debora MacKenzie, "Will a pandemic bring down civilisation?" *New Scientist*, April 2, 2008, www.newscientist.com/article/mg19826501-400-will-a-pandemic-bring-down-civilisation.

**183 But, as I...to fall:** Debora MacKenzie, "Why the demise of civilisation may be inevitable," *New Scientist*, April 2, 2008, www.newscientist.com/article/mg19826501-500-why-the-demise-of-civilisation-may-be-inevitable.

**184 The important...the other:** Thomas Homer-Dixon, "Complexity science," *Oxford Leadership Journal* 2, no. 1 (January 2011), homerdixon.com/complexity-science.

**184 The famous...outcome:** Edward N. Lorenz, "Predictability; does the flap of a butterfly's wings in Brazil set off a tornado in Texas?" American Association for the Advancement of Science, 139th meeting, December 29, 1972, eaps4.mit.edu/research/Lorenz/Butterfly_1972.pdf.

**188 The number...food to them:** Thin Lei Win and Kim Harrisberg, "Africa faces 'hunger pandemic' as coronavirus destroys jobs and fuels poverty," *Reuters*, April 24, 2020, www.reuters.com/article/us-health-coronavirus-africa-hunger-feat/africa-faces-hunger-pandemic-as-coronavirus-destroys-jobs-and-fuels-poverty-idUSKCN22629V.

**188 Vittoria...Algeria:** Marius Gilbert, et al., "Preparedness and vulnerability of African countries against importations of COVID-19: a modelling study," *The Lancet* 395, no. 10227 (March 2020): 871–77, doi.org/10.1016/s0140-6736(20)30411-6.

**191 The 2018 UK...Spanish flu:** Scientific Pandemic Influenza Group on Modelling, "SPI-M Modelling Summary," November 2018, assets.publishing.service.gov.uk/government/uploads/system/uploads/attachment_data/file/756738/SPI-M_modelling_summary_final.pdf.

**191 It acknowledges...unlikely:** Civil Contingencies Secretariat (UK), "Preparing for pandemic influenza: guidance for local planners," July 2013, assets.publishing.service.gov.uk/government/uploads/system/uploads/attachment_data/file/225869/Pandemic_Influenza_LRF_Guidance.pdf.

**193 As virologist...anytime soon:** "Expert reaction to preprint on COVID-19 and patient-derived mutations," *Science Media Centre*, April 21,

2020, www.sciencemediacentre.org/expert-reaction-to-preprint-on-covid -19-and-patient-derived-mutations.

**193 As I write...the virus:** Bette Korber, et al., "Spike mutation pipeline reveals the emergence of a more transmissible form of SARS-CoV-2," *bioRxiv*, May 5, 2020, doi.org/10.1101/2020.04.29.069054.

**194 Andrew Read...vaccinated for Marek's:** Andrew F. Read, et al., "Imperfect vaccination can enhance the transmission of highly virulent pathogens," *PLoS Biology* 13, no. 7 (July 2015), doi.org/10.1371/journal.pbio.1002198.

**196 Jeremy Luban...epidemic:** William E. Diehl, et al., "Ebola Virus Glycoprotein with Increased Infectivity Dominated the 2013–2016 Epidemic," *Cell* 167, no. 4 (November 2016): 1088–1098.e6, doi.org/10.1016/j .cell.2016.10.014.

**196 Andrew Read...evolve:** Debora MacKenzie, "Ebola rapidly evolves to be more transmissible and deadlier," *New Scientist*, November 3, 2016, www.newscientist.com/article/2111311-ebola-rapidly-evolves-to-be-more -transmissible-and-deadlier.

**197 In 2015...selected for resistance:** Peter Kerr, et al., "Myxoma virus and the leporipoxviruses: an evolutionary paradigm," *Viruses* 7, no. 3 (March 2015): 1020–61, doi.org/10.3390/v7031020.

**200 A subsequent...a week:** Alan Mckinnon, "Life without trucks: the impact of a temporary disruption of road freight transport on a national economy," *Journal of Business Logistics* 27, no. 2 (May 2006): 227–50, doi.org/10.1002/j.2158-1592.2006.tb00224.x.

**200 During Covid-19...a problem:** Debora MacKenzie, "Will a pandemic bring down civilisation?"

**201 The current...shut down:** Department for Business, Energy, and Industrial Strategy, and Health and Safety Executive (UK government), "Guidance: preparing for and responding to energy emergencies," January 9, 2020, www.gov.uk/guidance/preparing-for-and-responding-to-energy -emergencies.

**201 The current...shut down:** Department of Energy and Climate Change (UK), "DECC approach to dealing with pandemic illness in the upstream energy sector," July 24, 2013, assets.publishing.service.gov.uk /government/uploads/system/uploads/attachment_data/file/48946 /Dealing_with_pandemic_illness_in_the_upstream_energy_sector.doc.

**201 The official…reading:** Cybersecurity and Infrastructure Security Agency (US Department of Homeland Security), "Guidance on the essential critical infrastructure workforce," April 24, 2020, www.cisa.gov /publication/guidance-essential-critical-infrastructure-workforce.

**202 A massive…illnesses:** The OpenSAFELY Collaborative, et al., "OpenSAFELY: factors associated with COVID-19-related hospital death in the linked electronic health records of 17 million adult NHS patients." medRxiv, May 7, 2020, doi.org/10.1101/2020.05.06.20092999.

**204 Even in this…contagion:** Debora MacKenzie, "We don't know how Covid-19 spread on the Diamond Princess cruise ship," *New Scientist*, February 20, 2020, www.newscientist.com/article/2234734-we-dont -know-how-covid-19-spread-on-the-diamond-princess-cruise-ship.

**205 UN Secretary-General…climate change:** BBC, "Coronavirus: lack of co-ordination let virus spread—UN's Guterres," Television newscast, Interview by Nick Bryant, May 1, 2020, www.bbc.com/news/av/world-us-canada -52496983/coronavirus-lack-of-co-ordination-let-virus-spread-un-s-guterres.

**205 But that…average incomes:** Shannon K. O'Neill, "How to pandemic -proof globalization," *Foreign Affairs,* April 1, 2020, www.foreignaffairs .com/articles/2020-04-01/how-pandemic-proof-globalization.

**206 In fact, shipping…costs:** Adele Berti, "The impact of Covid-19 on global shipping: part 1, system shock," *Ship Technology*, April 2, 2020, www.ship-technology.com/features/impact-of-covid-19-on-shipping.

**207 Homer-Dixon agrees…stable state:** Thomas Homer-Dixon, et al., "Synchronous failure: the emerging causal architecture of global crisis," *Ecology and Society* 20, no. 3 (2015), doi.org/10.5751/es-07681-200306.

## CHAPTER 8

**209 We've got…no choice:** Sara Frueh, "NAS annual meeting: experts discuss COVID-19 pandemic and science's response," The National Academies of Science and Engineering, April 27, 2020, www.nation alacademies.org/news/2020/04/nas-annual-meeting-experts-discuss -covid-19-pandemic-and-sciences-response.

**209 When written…opportunity:** John F. Kennedy, "Remarks at the Convocation of the United Negro College Fund, Indianapolis, Indiana, April 12, 1959," JFK Library, www.jfklibrary.org/archives/other-resources

/john-f-kennedy-speeches/indianapolis-in-19590412. The quote is slightly different in its other iteration from October 1960.

**210 Covid-19…November 2019:** Josephine Ma, "Coronavirus: China's first confirmed Covid-19 case traced back to November 17."

**211 Secrecy…Tufekci:** Zeynep Tufekci, "How the coronavirus revealed authoritarianism's fatal flaw," *The Atlantic*, February 22, 2020, www.the atlantic.com/technology/archive/2020/02/coronavirus-and-blindness -authoritarianism/606922.

**211 by January 20th…China:** James Kynge, Sun Yu, and Tom Hancock, "Coronavirus: the cost of China's public health cover-up."

**215 I wrote…funding:** Debora MacKenzie, "Can we afford not to track deadly viruses?" *New Scientist*, May 20, 1995, www.newscientist.com /article/mg14619780-300-can-we-afford-not-to-track-deadly-viruses.

**217 The US tried…failed:** Nicholas Kulish, Sarah Kliff, and Jessica Silver-Greenberg, "The U.S. tried to build a new fleet of ventilators. The mission failed," *New York Times*, March 29, 2020, www.nytimes.com/2020/03/29 /business/coronavirus-us-ventilator-shortage.html.

**218 certainly not…valid claim:** Kristian Andersen, "nCoV-2019 codon usage and reservoir (not snakes v2)," *Virological*, January 24, 2020, virologi cal.org/t/ncov-2019-codon-usage-and-reservoir-not-snakes-v2/339.

**221 Zhengli Shi…sequenced:** Jane Qiu, "How China's 'Bat Woman' Hunted Down Viruses from SARS to the New Coronavirus," *Scientific American*, April 27, 2020, www.scientificamerican.com/article/how-chinas-bat -woman-hunted-down-viruses-from-sars-to-the-new-coronavirus1.

**221 Covid-19 was not…so well:** Kristian G. Andersen, et al., "The proximal origin of SARS-CoV-2."

**222 The G20 group…and treatments:** G20, "G20 leaders' statement, extraordinary G20 leaders' summit statement on COVID-19," March 26, 2020, g20.org/en/media/Documents/G20_Extraordinary%20G20%20Leaders %E2%80%99%20Summit_Statement_EN%20(3).pdf.

**225 Yet many…hit:** Mike Stobbe, "Health official says US missed some chances to slow virus," *Associated Press*, May 1, 2020, apnews.com /a758f05f337736e93dd0c280deff9b10.

**225 The virus is…for war:** Gary P. Pisano, Raffaella Sadun, and Michele Zanini, "Lessons from Italy's response to coronavirus," *Harvard*

*Business Review,* March 27, 2020, hbr.org/2020/03/lessons-from-italys-response-to-coronavirus.

**227 Some experts…meltdown too:** Adam Tooze, "How coronavirus almost brought down the global financial system," *Guardian,* April 14, 2020, www.theguardian.com/business/2020/apr/14/how-coronavirus-almost-brought-down-the-global-financial-system.

**227 shut down…ever known:** Christopher J. Fettweis, "Unipolarity, hegemony, and the new peace," *Security Studies* 26, no. 3 (August 2017): 423–51, doi.org/10.1080/09636412.2017.1306394.

**229 Larry Gostin…to overcome it:** Lawrence Gostin and Sarah Wetter, "Two legal experts explain why the U.S. should not pull funding from the WHO amid COVID-19 pandemic."

**229 In May 2020…Covid-19:** BBC, "Coronavirus: lack of co-ordination let virus spread—UN's Guterres."

**230 in 2014, the WHO's…investment:** Debora MacKenzie, "World must get ready now for the next big health threat."

**231 Hierarchies are already…experts:** Anne-Marie Slaughter, *The Chessboard and the Web: Strategies of Connection in a Networked World* (New Haven, CT: Yale UP, 2017).

**231 Bill Gates…outbreaks:** Bill Gates, "Bill Gates on how to fight future pandemics," The Economist, April 23, 2020, www.economist.com/by-invitation/2020/04/23/bill-gates-on-how-to-fight-future-pandemics.

**234 chemical disarming…weakening:** Debora MacKenzie, "US may respond after chemical weapons attack in Syria," *New Scientist,* April 11, 2018, www.newscientist.com/article/mg23831733-600-us-may-respond-after-chemical-weapons-attack-in-syria.

**234 Treaty countries…of refusal:** Organisation for the Prohibition of Chemical Weapons, "Chemical Weapons Convention," September 27, 2005 (revised), www.opcw.org/chemical-weapons-convention.

**234 except the US…refuse:** Jonathan B. Tucker, "The chemical weapons convention: has it enhanced U.S. security?" *Arms Control Today,* April 2001, www.armscontrol.org/act/2001-04/features/chemical-weapons-convention-enhanced-us-security.

**235 There is already…hard enough:** World Health Organization, "Global Polio Eradication Initiative," polioeradication.org.

**236 In 2004...Pathogens Treaty:** Debora MacKenzie, "The great flu cover-up," *New Scientist*, January 31, 2004, www.newscientist.com /article/mg18124320-200-the-great-flu-cover-up.

**237 by 2030...need help:** Debora MacKenzie, "Chasing deadly viruses for a living," *New Scientist*, July 4, 2012, www.newscientist.com/article /mg21528722-100-chasing-deadly-viruses-for-a-living.

**238 UN Secretary...economy:** António Guterres, "Secretary-General's remarks at G-20 virtual summit on the COVID-19 pandemic," United Nations, March 26, 2020, www.un.org/sg/en/content/sg/statement/2020 -03-26/secretary-generals-remarks-g-20-virtual-summit-the-covid-19 -pandemic.

**238 British columnist...beginning:** Tim Walker, Twitter Post, March 28, 2020, 2:03 PM, twitter.com/ThatTimWalker/status/1243961867116204032.

**238 Jonathan Weigel...all of us:** Maitreesh Ghatak, Xavier Jaravel, and Jonathan Weigel, "The world has a $2.5 trillion problem. Here's how to solve it," *New York Times*, April 20, 2020, www.nytimes.com/2020/04/20 /opinion/coronavirus-economy-bailout.html.

**239 Besides our biological...infection:** Mark Schaller, "The behavioural immune system and the psychology of human sociality," *Philosophical Transactions of the Royal Society B: Biological Sciences* 366, no. 1583 (December 2011): 3418–26, doi.org/10.1098/rstb.2011.0029.

**239 People with...historically:** Kathleen McAuliffe, "Liberals and conservatives react in wildly different ways to repulsive pictures," *The Atlantic*, March 2019, www.theatlantic.com/magazine/archive/2019/03/the-yuck -factor/580465.

**239 People with...historically:** Corinne J. Brenner and Yoel Inbar, "Disgust sensitivity predicts political ideology and policy attitudes in the Netherlands," *European Journal of Social Psychology* 45, no. 1 (November 2014): 27–38, doi.org/10.1002/ejsp.2072.

**239 People with...historically:** Corey L. Fincher, et al., "Pathogen prevalence predicts human cross-cultural variability in individualism /collectivism," *Proceedings of the Royal Society B: Biological Sciences* 275, no. 1640 (February 2008): 1279–85, doi.org/10.1098/rspb.2008.0094.

**240 Researchers have...measured:** Debora MacKenzie, "How your personality predicts your attitudes towards Brexit," *New Scientist*, July 9, 2018,

www.newscientist.com/article/2173681-how-your-personality-predicts
-your-attitudes-towards-brexit.

240 **Cambridge...as well:** Leor Zmigrod, et al., "The psychological
and socio-political consequences of infectious diseases," *PxyArXiv Pre-
prints* (April 11, 2020), doi.org/10.31234/osf.io/84qcm.

240 **Mark Schaller...Canadians:** Alec T. Beall, et al., "Infections and
elections." *Psychological Science* 27, no. 5 (March 14, 2016): 595–605. doi
.org/10.1177/0956797616628861.

241 **The Centre...governments:** Arnstein Aassve, Guido Alfani, Fran-
cesco Gandolfi, and Marco Le Moglie, "Pandemics and social capital: from
the Spanish flu of 1918-19 to COVID-19," *VoxEU*, March 22, 2020, voxeu
.org/article/pandemics-and-social-capital.

241 **As a presidential...disease:** Philip Bump, "Donald Trump's
lengthy and curious defense of his immigrant comments, annotated,"
*Washington Post*, July 6, 2015, www.washingtonpost.com/news/the-fix
/wp/2015/07/06/donald-trumps-lengthy-and-curious-defense-of-his
-immigrant-comments-annotated.

241 **In February...in waiting:** Tierra Smiley Evans, et al., "Synergistic
China–US ecological research is essential for global emerging infectious
disease preparedness," *EcoHealth* 17, no. 1 (March 2020): 160–73, doi
.org/10.1007/s10393-020-01471-2.

242 **American author...altruism:** Rebecca Solnit, *A Paradise Built in
Hell* (New York, NY: Viking, 2009).

243 **Both the US...virus:** Steven Lee Myers, "China spins tale that
the U.S. Army started the coronavirus epidemic," *New York Times*, March
13, 2020, www.nytimes.com/2020/03/13/world/asia/coronavirus-china
-conspiracy-theory.html.

243 **Some American...January:** Marc A. Thiessen, "China should be
legally liable for the pandemic damage it has done," *Washington Post*, April
9, 2020, www.washingtonpost.com/opinions/2020/04/09/china-should
-be-legally-liable-pandemic-damage-it-has-done.

243 **In April...compelling:** "Statement: Saving Lives in America, China,
and Around the World," signed Madeleine Albright, et al., UC San Diego
21 Century China Center, April 3, 2020, china.ucsd.edu/_files/statement
/covid-19-pandemic-statement.pdf.

243 **In May…contribute to that:** Laurens Cerulus, "Ursula von der Leyen backs probe into how coronavirus emerged," *Politico EU,* May 1, 2020, politico.eu/article/von-der-leyen-backs-probe-into-how-coronavirus -emerged.

244 **Jeremy Farrar…cohesive world:** The version quoted here is a slightly refined version Farrar tweeted the day after the talk: Jeremy Farrar, Twitter Post, April 26, 2020, 6:26 AM, twitter.com/JeremyFarrar /status/1254356097470738432. For the original speech: Jeremy Farrar, "COVID-19 Update," Panel discussion, National Academy of Sciences 157th Annual Meeting, April 25, 2020, online, www.nasonline.org/about -nas/events/annual-meeting/nas157/covid19-update.html.